Anatomy, Physiology and Hygiene

Anatomy, Physiology and Hygiene

Janet K. Raeburn MA (Cantab.), Headmistress, Westcliff High School

H. A. Raeburn MD, FRCP (Edin.), DPH

In collaboration with

Hilda M. Gration SRN, SCM, Diploma in Nursing (University of London)

THIRD EDITION

John Murray Fifty Albemarle Street London

First edition 1940
Reprinted 1944, 1947, 1950, 1954, 1956
Second edition 1957
Reprinted 1958, 1960, 1962, 1964
1965 (revised), 1966, 1967
Third edition 1969

Printed in Great Britain by
Cox & Wyman Ltd, London, Fakenham and Reading

7195 1138 0 boards
7195 1906 3 paperback

Contents

Part two

Hygiene

Preface to the third edition

In producing this new edition of Anatomy, Physiology and Hygiene, our purpose has been to bring the book up to date without altering its character and its emphasis on the relationships between structure and function.

A simplified account of some aspects of Molecular Biology and the significance of the nucleic acids DNA and RNA will be found in the Appendix to Part One. For this chapter we are indebted to Mr H. M. Thomas, Senior Lecturer in Biology and Health Education at Furzedown College of Education, who has worked on this subject with his students.

Our thanks are also due to Mrs Judith Rich, who has redrawn with care and patience the majority of the diagrams; to Dr R. Harvey-Kelly, who has read the text and given advice about the modern work in Physiology; and above all to the staff of John Murray's educational department for their determination and patient zeal in initiating and carrying through this revision.

J. K. R.
H. A. R.

Part one

Anatomy and Physiology

1 Introduction

The larger animals, including human beings, consist of microscopic units, known as cells, which make up the 'living matter' or protoplasm, together with non-living materials which are products of some of these cells.

The cellular structure of the body can be illustrated by examining under the microscope scrapings of the inside of the cheek. The cellular units can be clearly distinguished, the central nucleus made out and probably granules in the non-nuclear part of the protoplasm, known as the cytoplasm.

Chemical analysis shows that the chief elements of which the cell is composed are hydrogen, oxygen and nitrogen together with sulphur and phosphorus.

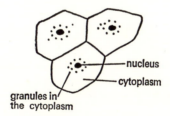

nucleus
cytoplasm
granules in
the cytoplasm

Fig. 1. Epithelial cells from the inside of the cheek.
(Highly magnified.)

The detailed structure of the cell is a fascinating and very important field of study and great advances have been made in it in recent years.

It is possible to know a great deal about how the body works, without going into the fine structure of living tissue, but the real basic questions (such as how the correct set of 'instructions' or 'programme' is passed on as cells multiply by dividing) can be answered only in terms of the structure and functioning of the cells – a subject which is dealt with in the Appendix to Part One.

The general characteristics of living matter may be shown in a relatively simple way by examining a small one-celled animal such as the Amoeba, which consists of a small mass of protoplasm about 0·1 mm in diameter.

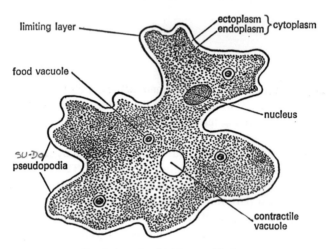

Fig. 2. Amoeba (highly magnified).

Characteristics of life as shown by Amoeba

(1) Movement. The shape of the Amoeba is continually changing as the animal puts out in all directions projections of protoplasm called false feet or pseudopodia. The rest of the protoplasm follows the longest projection, and so the animal moves on from place to place.

(2) Irritability. The animal does not move about indiscriminately, but is sensitive to changes in its environment, in other words it is irritable. It moves away from harmful chemicals in the water, away from a very bright light, but towards food particles and oxygen.

(3) Respiration. The Amoeba requires energy for all its life processes. It obtains this energy by oxidizing or burning food. For this to go on it must have oxygen. This it takes in in solution throughout the entire surface, and carbon dioxide, the waste product of the process, is given out in solution throughout the entire surface. The burning of food is spoken of as internal respiration. The necessary interchange of gases between the animal and the environment is called external respiration.

(4) *Nutrition.* When the Amoeba comes across particles of organic matter in the water, it puts out two pseudopodia, one on each side of the particle. The two pseudopodia meet, and the particle comes to lie in a space in the protoplasm called the food vacuole.

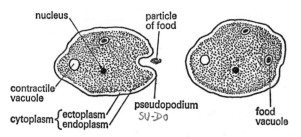

Fig. 3. Amoeba taking in or ingesting a food particle.

Before the food particle can be utilized, it has to be converted into soluble substances which can be absorbed into the protoplasm, i.e. it has to be digested. Once absorbed, the protoplasm can utilize the food, using it either as a source of energy or for building up new protoplasm.

Any food which cannot be digested inside the vacuole the animal leaves behind in the water, and the food vacuole disappears.

(5) *Growth.* The new protoplasm built up either replaces that broken down during the 'wear and tear' of the cell, or adds to the cell, i.e. the cell grows.

(6) *Excretion.* During the activities of the cell certain waste substances are produced. These, if they remained inside the cell body, would be harmful. They are therefore passed to the outside, and this process is known as excretion.

In addition to carbon dioxide, which has already been mentioned, a certain amount of protoplasm is broken down during the ordinary 'wear and tear'. Protoplasm contains nitrogen, and so this waste is called nitrogenous waste. The animal excretes it in solution, mostly in the form of ammonia, either throughout the entire surface, or by an apparatus known as the contractile vacuole. The latter is a space in the protoplasm into which water with substances in solution is continually being passed. It contracts from time to time, and forces its contents to the outside. In this way the animal gets rid of the surplus water which it accumulates as a result of osmosis.

(7) Reproduction. When the Amoeba cell reaches the adult size it divides into two, the nucleus dividing before the rest of the protoplasm. In this way two 'daughter' Amoebae are formed in place of the original cell.

These new cells grow, and when they reach the adult size in their turn divide.

Although millions of Amoebae are killed off by adverse conditions, there is nothing in the life history of this animal corresponding to natural death. The protoplasm is immortal, being handed on from one generation to the next.

Multicellular animals

The Amoeba consists of one cell, which is capable of carrying out all the vital functions. But the majority of animal species are multicellular and are divided up into the soma or body cells, and the reproductive cells which are protected by the soma.

Amongst the somatic cells there is a definite division of labour, cells or groups of cells having their own particular function. The beginning of this division of labour may be seen by studying a small water animal called Hydra.

Hydra. This animal consists of a sac about 6–7 mm long. It is attached by its closed end to water weed.

The wall of the sac is made up of two layers of cells, the outer layer or ectoderm, and the inner layer or endoderm. Between the two cellular layers there is a structureless layer of jelly called the mesogloea. Around the mouth of the sac is a ring of tentacles.

The ectoderm is made up of different kinds of cells modified for the functions which they perform:

(1) *The musculo-epithelial cells.* These protect the animal. They also help the animal to shorten by the contraction of their muscular tails (see Fig. 6).

(2) *The cnidoblasts.* These help the animal to capture its food. A short hair, the cnidocil, projects from each cell, and when this is touched a long poisonous dart is shot out by pressure from an oval sac, and so the prey is trapped. The prey is then pushed by the tentacles in through the mouth.

(3) *Nerve cells.* These are highly sensitive. They are extended at the corners into long processes which come in contact with one another to

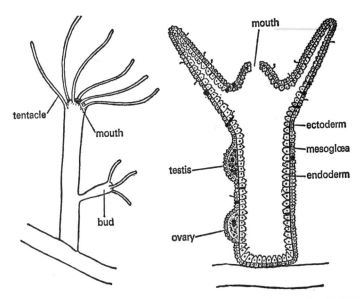

Fig. 4. Hydra.   Fig. 5. Longitudinal section of Hydra.

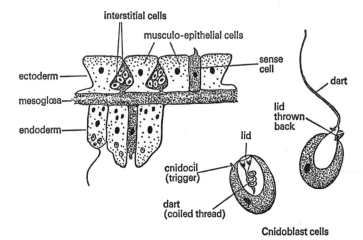

Fig. 6. Cells of Hydra.

form a nerve net, lying on the surface of the mesogloea. Thus if one nerve cell is stimulated by changes in the surroundings, an impulse may be conveyed over the entire animal.

(4) *Interstitial cells.* These lie in patches between some of the musculo-epithelial cells. They can give rise to new ectoderm cells, and also to reproductive cells.

The endoderm is made of cells of two kinds, some large and some small. The large cells put out pseudopodia similar to the Amoeba and are able to engulf food particles from the sac of the animal. The smaller cells secrete a digestive fluid which passes into the sac and carries on a certain amount of digestion there.

Some of the endoderm cells possess flagella, hair-like processes which by their beating keep up a circulation of water within the sac.

The endoderm cells similar to the musculo-epithelial cells of the ectoderm have muscular tails. These run round the animal, and when they contract the animal becomes longer and thinner.

At times some of the interstitial cells of the ectoderm divide up, and so produce small swellings on the side of the animal. A swelling may become a male reproductive organ or testis, containing male cells or sperms, or it may become a female reproductive organ or ovary, containing one large egg cell or ovum, which has developed at the expense of the other cells.

The sperms escape into the water, then one sperm enters the egg cell and fertilization takes place. The fertilized egg begins to divide up to form an embryo, which becomes surrounded by a thick wall. The embryo then falls off the parent animal in a state fit to resist adverse conditions.

Sometimes Hydra reproduces by forming a bud (see Fig. 4), which breaks away from the parent ready to start a separate existence.

Thus in the Hydra there is a definite division into soma or body cells which protect the reproductive cells. Within the soma there is a certain amount of cell differentiation, although in this animal specialization is not advanced, and there are no cells specially developed for excretion and respiration. An interchange of material between the cells and the surrounding water goes on throughout the entire surface.

In the bodies of the higher animals, which include man, division of labour is carried further. The higher animals consist of masses of cells which are not arranged haphazardly, cells having the same structure and function being grouped together.

For example, in man, movement is effected by cells which are long delicate structures and have the power of contraction and elongation. These cells, known as muscle fibres, are grouped together to form powerful structures called muscles, which move and give support to the body.

Part One of this book describes the Anatomy and Physiology of the human body. 'Anatomy' is the study of the structure and the relative positions of the various parts of the body. 'Physiology' is the study of the functions of the organs and their co-ordination for the smooth working of the body as a whole.

Man

Modern man or *Homo sapiens* made his appearance about 30,000 B.C. although other species of man had lived on the earth for over 350,000 years before that.

Man belongs to the order of mammals known as the Primates. He is distinguished from the other primates by the following physical characteristics:

(1) Erect posture and bipedal gait with limb structure suitable for this. The hind limbs or legs are longer than the front limbs and support the body. The foot, although not able to grasp objects, is adapted for walking and running with its high arch and big toe in line with the others.

(2) Hands free, with an opposable thumb and first finger making it possible to manipulate objects.

(3) Head fixed in such a way on the spinal column as to enable him to look straight ahead when standing upright with the face in the vertical plane. He has a prominent nose.

(4) Well developed brain.

Primitive man developed verbal communication, a primitive form of art and, most important, the faculty of conceptual thought. Unlike many other primates, he was able not only to use tools, but also to design them, having in mind the idea of their possible use in the future for the control of his environment. Having reached the stage we call *Homo sapiens*, man began to wonder about what happened after death and he buried his dead with a supply of food and tools for use in another world. He developed a specialized social structure.

All men living today are members of the single species, *Homo sapiens*, and can interbreed freely. There are, however, different racial types. For a description of these, students are referred to books on human biology.

PRACTICAL

1 Carefully scrape the inside of the cheek with the blunt end of a scalpel. Mount the scrapings in saliva on a glass slide and cover with a cover-slip. Examine under the microscope.

2 Amoeba

   (*a*) Mount a drop of water containing Amoebae on a slide. Examine under the low power. Observe the movements of an Amoeba.

   (*b*) Examine a prepared slide of an Amoeba under the high power.

3 Hydra

   (*a*) Examine the living animal with a hand lens in water in a watch-glass.

   (*b*) Mount a Hydra in water on a slide. Cover with a cover-slip supported with thin strips of paper. Examine under the low power.

   (*c*) Examine a tentacle under the high power. Irritate with dilute acetic acid and watch. The nematocyst threads will be shot out.

   (*d*) Examine prepared slides of a transverse section and a longitudinal section of Hydra. Note ectoderm and endoderm cells.

2 The body

The millions of cells which make up multicellular animals, including man's, are not all alike. They differ in many ways, such as size, shape, composition and function.

Cells are grouped together into tissues, a tissue being composed of a number of cells identical in structure and function and held together by ground substance or matrix.

In the body there are five basic tissues:

(1) Epithelial.
(2) Connective.
(3) Muscular.   EP—I—THE—LI—UM
(4) Nerve.
(5) Fluid.

Epithelial tissue

Epithelial tissue or 'epithelium' is formed of cells specialized to cover the surface of the body and to line internal cavities such as the mouth and stomach. The epithelial tissue lining blood vessels and lymph vessels (see page 86) is called endothelium.

The cells of an epithelium touch one another, there being very little or no intercellular matrix. They are situated on a basement membrane. There are two main types of epithelium:

(i) *Simple,* in which there is only one layer of cells.
(ii) *Compound,* where there are several layers of cells.

Simple epithelium. There are various types of simple epithelium classified according to the shape of the individual cells, the form of the free border and function.

(1) *Squamous or pavement epithelium,* in which the cells are flattened and scale-like, and arranged edge to edge, e.g. the epithelium lining the alveoli of lung.

(2) *Columnar epithelium,* in which the cells are column shaped, their long axis being at right angles to the basement membrane, e.g. the epithelium lining the inside of the stomach and intestines.

(3) *Cubical epithelium,* in which the cells are nearly square, e.g. the epithelium found inside the kidney.

(4) *Ciliated epithelium,* in which short actively waving threads called cilia are attached to the free surface of the cells. These cells are generally columnar in shape, e.g. the epithelium lining the windpipe or trachea.

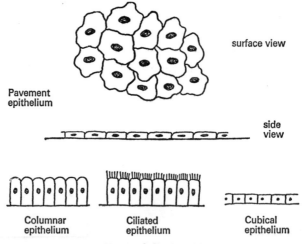

surface view

Pavement epithelium

side view

Columnar epithelium   Ciliated epithelium   Cubical epithelium

Fig. 7. Types of simple epithelium.

(5) *Glandular epithelium.* Many cells are set aside in all parts of the body for giving out or secreting useful substances. Sometimes the cells are isolated and lie amongst other epithelial cells, e.g. goblet cells which lie amongst the columnar epithelial cells lining the gut, but in most cases they are grouped together to form a glandular epithelium which has sunk in from the surface to form glands of different shapes:

(i) Test-tube shaped glands lined with secretory epithelium, e.g. the gastric glands in the stomach wall.

(ii) Coiled glands, longer than (i) and coiled at the end. Only the coiled part is secretory, the rest is lined with non-secretory cells and forms the duct, e.g. sweat gland.

(iii) Compound glands, which have many branches. In an alveolar compound or racemose gland, e.g. salivary glands, pancreas, the ends of the branches become enlarged into bulbous chambers or alveoli, which are surrounded by secretory cells. The rest of the branches are lined with non-secretory cells.

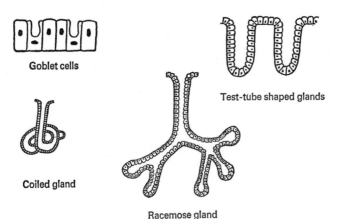

Goblet cells

Test-tube shaped glands

Coiled gland

Racemose gland

Fig. 8. Types of glandular epithelium.

(iv) Ductless glands. These glands, e.g. thyroid, consist of masses of epithelial cells with no duct leading away. The secretions are poured directly into the bloodstream.

(6) *Sensory epithelium.* In certain parts of the body epithelial cells are sensitive to certain stimuli, for example in the nose, eye and ear. These cells will be described in Chapter 17.

Compound epithelium. Compound or stratified epithelium forms the outer covering of the skin and lines the mouth, pharynx, oesophagus and vagina. It consists of layers of cells, the deeper cells having a distinct shape, and protoplasmic contents, but nearer the surface the cells gradually become flatter and scale-like and in the skin lose their living contents.

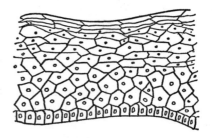

Fig. 9. Compound epithelium.

Connective tissue

Connective tissue binds the various parts of the body together, and helps in the support of the body. It includes bony tissue highly specialized for support and is characterized by a large amount of intercellular matrix produced by the living cells, which after they have done this are of very little importance. There are various kinds of connective tissue:

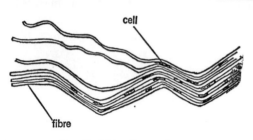

Fig. 10. White fibrous tissue.

White fibrous tissue. This is a tough inelastic tissue made up of very fine white wavy fibres collected together into bundles, in which they run parallel. These fibres, which do not branch, form the matrix; the cells can sometimes be seen in between the fibres.

White fibrous tissue is found pure in tendons which attach muscles to bone.

Yellow elastic tissue. This is an elastic tissue made up of yellow fibres which, unlike the white fibres, branch and join up again. Cells are sometimes seen between the fibres.

✳ A-RE-O-LAR.

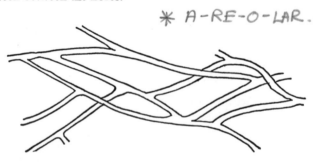

Fig. 11. Yellow elastic tissue.

Yellow elastic tissue is the chief (but not the only) component of ligaments, which help to form joints.

✳ *Areolar tissue.* In many parts the white fibres and yellow fibres are found mixed together with various amounts of non-fibrous ground substance forming areolar tissue. This tissue is found, for example, under

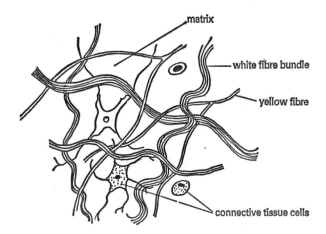

Fig. 12. Areolar tissue.

the skin, binding it to the deeper structures; it also surrounds the various internal organs and keeps them in place.

Adipose tissue is the fatty tissue of the body. It is connective tissue with few fibres and the cells swollen with fat droplets. Fatty tissue acts as a protection to the organs close to it, or the fat acts as an emergency food reserve.

Cartilage, or gristle, consists of a tough matrix of organic matter. In this matrix are small cavities containing at first one cartilage cell. This cell may divide up to form two, and then four cells. These cells are continually producing fresh matrix which they lay down around and between themselves, so becoming farther and farther apart. Then they in their turn divide.

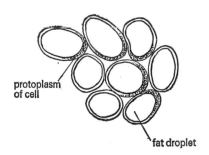

Fig. 13. Adipose tissue.

This type of cartilage described is hyaline cartilage and is found for example in the tracheal wall, at the end of ribs, where it is known as costal cartilage, and at the end of long bones, where it is known as articular cartilage. In fibro-cartilage the matrix contains white fibres, and in elasto-cartilage the matrix contains yellow fibres. Fibro-cartilage

FIBRO | CARTILAGE

forms, for example, the intervertebral discs, which are pads of cartilage between individual vertebrae, the bones which comprise the backbone. Elasto-cartilage is found, for example, in the epiglottis.

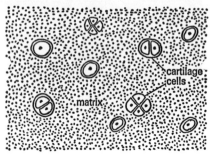

Fig. 14. Hyaline cartilage.

Bone is very firm connective tissue, the bone cells being separated by a matrix of an organic material containing lime salts, chiefly calcium phosphate. This matrix gives bone its hardness and strength.

There are two types of bone: compact bone and spongy or cancellous bone.

If a section of a compact bone is examined under the microscope it will be seen to be made up of groups of concentric layers of the bony

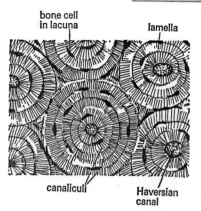

Fig. 15. A section of compact bone.

LA-MEL-A

matrix called lamellae, each surrounding a central canal. This is called the Haversian canal, and contains blood vessels and nerves. In between the lamellae lie small cavities, the lacunae, in which the bone cells are situated.

LA-KU-NA

The lacunae are united with one another and with the Haversian canal by small canaliculi, which run through the lamellae. Branches of bone cells run into these channels.

The Haversian canal surrounded by its lamellae is called a Haversian system.

(SKULL BONE)

In spongy bone struts of matrix called trabeculae surround large spaces called Haversian spaces. These are filled with red marrow.

Muscular tissue

Muscular tissue, which has the power of contraction, is made up of cells which are elongated and called fibres. There are three main types of muscular tissue: (1) striped, skeletal or voluntary, (2) unstriped, plain, visceral or involuntary, (3) cardiac.

(1) *Striped, skeletal or voluntary muscular tissue.* This tissue builds up all muscles which are under control of the will, hence the name voluntary. It includes most of the 'meat' or 'flesh' of an animal.

A striped muscle fibre is long and narrow. It contains protoplasm called saroplasm, which is surrounded by a wall of firmer protoplasm called the sarcolemma. Under this wall lie the nuclei, a number in each fibre.

The sarcoplasm contains long contractile fibrils which consist of alternating segments of light and dark material. These give the fibre a striated appearance.

(2) *Unstriped, plain, visceral or involuntary muscular tissue.* This builds up sheets of muscles which are not under the control of the will, for example, those found in the wall of the gut and in the walls of the blood vessels. Unstriped muscles show a slow contraction which they are capable of sustaining even though separated from the central nervous system. They are very sensitive to mechanical stimulation such as stretching.

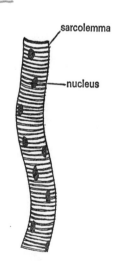

sarcolemma

nucleus

Fig. 16. Striped muscle fibre.

The unstriped muscle fibres are spindle shaped with tapering ends. Each fibre contains only one nucleus in a central position, and although sarcoplasm contains long contractile fibrils there are no clear cross striations.

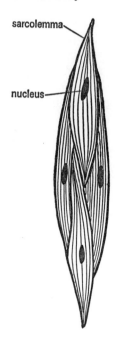

sarcolemma

nucleus

Fig. 17. Plain muscle
fibres.

(3) *Cardiac muscular tissue.* Cardiac muscle fibres are only found in the wall of the heart. They have no sarcolemma and only faint cross striations. They are not properly divided into separate units, and they branch and join up with their neighbours so that a kind of network is formed.

Nervous tissue

Nervous tissue consists of nervous cells or neurones with supporting tissue called neuroglia. A neurone has a cell body (sometimes called the cell) with one or more dendrons, processes with many branches, which conduct nerve messages or impulses toward the cell body, and one axon, a process which conducts the impulse away from the cell body. The axon, which is a slender thread, sometimes as long as three of four feet, branches near its termination. The cell body contains a large nucleus and has a very granular cytoplasm. In a sensory neurone, which conveys impulses from the outside world to the central nervous system, there is only one dendron; in a motor neurone, which conveys impulses from the central nervous system to an organ which reacts, and in a connector neurone between the sensory and motor neurones, there are a number of dendrons.

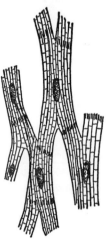

Fig. 18. Cardiac
muscle fibres.

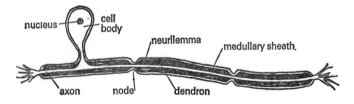

nucleus — cell body

neurilemma — medullary sheath.

axon   node   dendron

Fig. 19. A sensory neurone.

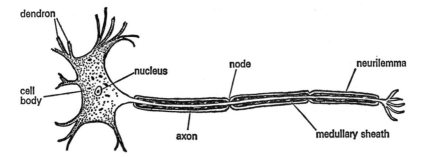

Fig. 20. A motor neurone.

The axon and, in the sensory neurone, the dendron, are covered on the outside with a tough nucleated coat, the neurilemma, and in the majority of neurones between the neurilemma and axon there is a medullary sheath of fatty material which gives the axon its white appearance. This sheath is interrupted at certain points called nodes, where the neurilemma sinks down and comes in contact with the axis cylinder.

The relationship between a sensory and a motor neurone in a nerve circuit may be seen in Fig. 92, page 159.

Nerve cells have a great power of regeneration if the cell body is un-injured, a new axon growing out from the cell body.

Fluid tissue

In fluid tissue which is often classed as a form of connective tissue, the intercellular matrix is liquid, and in it the tissue cells move about, e.g. blood, lymph. These will be described in Chapter 6 and Chapter 9.

Organs

The various tissues are grouped together to form the organs of the body. An organ is composed of at least two tissues closely associated in position and functions.

Two organs, a skeletal muscle and a long bone, can serve as illustrations.

Muscle. In a muscle the striated fibres, each surrounded with delicate connective tissue called endomysium, are collected together into bundles or fasciculi. Each fasciculus is surrounded by a sheath known as the

FAS-1-

perimysium, and several fasciculi are collected together to form a muscle. On the outside of a muscle is a dense elastic membrane called the epimysium. This is continuous with the perimysium surrounding each fasciculus.

Small blood vessels called capillaries ramify between the bundles and fibres, and there are many nerve endings lying on the fibres.

The muscle is attached to bone by tendons.

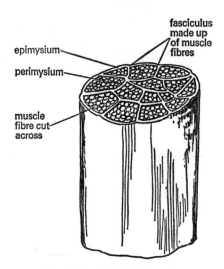

Fig. 21. Part of striped muscle.

Long bone. A long bone of the limbs consists of a central part, the shaft or diaphysis, and two ends, the epiphyses.

If a long bone is sawn along its length it will be seen that the shaft consists of compact bone surrounding a tubular cavity, the marrow cavity containing yellow marrow cells full of fat. The ends consist of cancellous bone, the Haversian spaces contain red marrow cells, which form new red blood corpuscles. Here there is no marrow cavity, and the spongy bone is covered with a thin layer of compact bone.

The Haversian canals in the compact part of the long bone run along the axis and branch and unite. These canals contain vessels and nerves which unite with those in the periosteum, a fibrous tissue membrane covering the outside of the bone. This membrane has a protective and nutrient function and is important in the growth and repair of bone. It is not present over the ends of the bone, where there is hyaline cartilage covering the articulating surfaces (articular cartilage).

In addition to the periosteal vessels, nutrient vessels enter and leave the marrow cavity, via the nutrient foramina, small holes which are well marked in long bones.

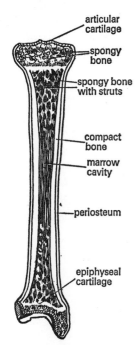

Development of bone. Bones develop in two ways. Long bones develop in bars of cartilage, for example the thigh bone in the developing individual before birth exists as a bar of cartilage. Many flat bones develop in membranes.

Bone formation always starts at certain definite centres which are constant in number and position for any one bone.

In a long bone there is one primary centre which appears before birth. It is in the centre and produces the shaft. After birth two secondary centres arise, one at either end, and they produce the epiphyses. The outside of these or the articular cartilage does not become ossified.

Fig. 22. Longitudinal section of tibia (shin bone).

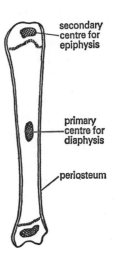

Fig. 23. Centres of ossification in a long bone.

Between an epiphysis and a bone shaft there is a piece of growing cartilage (the epiphyseal cartilage), and as fast as this cartilage grows it is turned into bone and so the bone continues to grow in length. At the age of about twenty-five in human beings, the entire plate becomes ossified.

Growth in thickness during juvenile life takes place by deposition of bone under the periosteum.

The process of bone formation is always essentially the same. It takes place in three stages. First of all cartilage cells enlarge and arrange themselves in rows, and calcium salts are laid down in the matrix. At the same time certain bone-forming cells, the osteoblasts, lay down a layer of bone and fibrous material under the periosteum. Some of the osteoblasts are included in this layer and become the bone cells.

After that cartilage-destroying cells, the osteoclasts, produced from the developing bone, invade the cartilage and eat away the matrix. These osteoclasts are followed by osteoblasts, which form bone in the spaces which have been eaten out. Blood capillaries accompany the osteoblasts, and ramify in the spaces.

At first bone is of the spongy type, but as the osteoblasts go on laying down bone matrix, it is laid down in concentric lamellae.

Other organs will be described throughout the course of this book.

System

A body system is composed of a number of organs concerned with similar functions. The following is a list of the most important systems found in the body:

(1) Skeletal and connective system.
(2) Muscular system.
(3) Respiratory system.
(4) Circulatory system.
(5) Digestive and glandular system.
(6) Excretory system.
(7) Nervous system.
(8) Reproductive system.

The connection between a cell and a system may be shown diagrammatically:

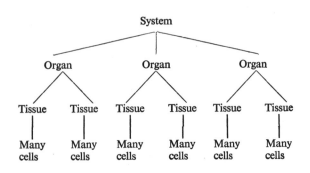

e.g.

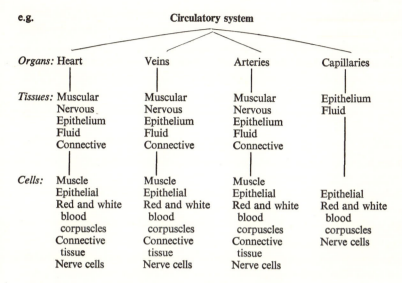

Circulatory system

	Heart	Veins	Arteries	Capillaries
Organs:	Heart	Veins	Arteries	Capillaries
Tissues:	Muscular Nervous Epithelium Fluid Connective	Muscular Nervous Epithelium Fluid Connective	Muscular Nervous Epithelium Fluid Connective	Epithelium Fluid

Cells:	Muscle
	Epithelial
	Red and white
	  blood
	  corpuscles
	Connective
	  tissue
	Nerve cells
		Muscle
		Epithelial
		Red and white
		  blood
		  corpuscles
		Connective
		  tissue
		Nerve cells
			Muscle
			Epithelial
			Red and white
			  blood
			  corpuscles
			Connective
			  tissue
			Nerve cells
				Epithelial
				Red and white
				  blood
				  corpuscles
				Nerve cells

Reticulo-endothelial system (eg Spleen)

Another system which will be mentioned throughout the course of this book is the reticulo-endothelial system. It consists of isolated endothelial cells, which are capable of removing certain substances which cannot be dealt with by the blood. For example, certain cells in the organ known as the spleen remove worn-out red blood corpuscles.

When necessary the cells of this system are able to free themselves and enter the bloodstream. They then show amoeboid motion.

PRACTICAL

For the work on fresh material no elaborate staining is necessary. Specimens may be stained with haematoxylin to bring out the nuclei. They should be examined microscopically.

Epithelium

1 Squamous epithelium. Scrape the inside of the cheek gently with the handle of a scalpel. Mount in saliva and examine.
2 Examine a prepared slide of T.S. of skin. Note stratified epithelium and sweat glands.
3 Columnar epithelium. Examine T.S. of the stomach and test-tube shaped glands.
4 Racemose glands. Examine T.S. of salivary gland or pancreas.

24 The body

Connective tissue

1 Areolar tissue. Mount in salt solution membranes of connective tissue between skin and muscle of rabbit.
2 Ligament. Examine fragment of sheep's ligament teased in salt solution.
3 Cartilage. Examine piece of upper or lower end of frog's sternum pressed out on a slide in salt solution.
4 Cartilage. Examine T.S. of trachea and notice incomplete rings of cartilage.
5 Bone. Examine T.S. of compact bone.
6 Bone. Examine slide to show ossification of bone.
7 Bone. Examine long bone sawn along its length.
8 Bone. Place bone in dilute hydrochloric acid in order to remove the inorganic matter. Note the result.
9 Bone. Roast bone in crucible in order to remove the organic matter. Note the result.

Muscle

1 Striped muscle. Dissect out gastrocnemius muscle of rabbit or frog. Show connection with sciatic nerve and the connection of the sciatic nerve with central nervous system.
2 Striped muscle. Tease out fragment of muscle. Mount in salt solution.
3 Examine slides of striped muscle fibres and muscles.
4 Tease out some sheep's cardiac muscle in salt solution on slide.
5 Examine T.S. of stomach to show unstriped muscle fibres cut in three directions.

3 The body as a unit

Man is a backboned animal. The backbone or spinal column consists of a number of small bones called vertebrae. Through it runs the spinal canal, which contains the spinal cord, part of the nervous system. The spinal cord gives off paired spinal nerves which pass out between the vertebrae.

The spinal cord expands in the head to form the brain, which is protected by the skull, a bony case attached to the first vertebra. The cavity inside the skull in which the brain is lodged is called the cranial cavity. The brain gives off cranial nerves which pass out through holes or foramina in the skull. The rest of the head is made up of the face.

From the vertebrae in the chest region ribs pass sideways and forwards to the front, where they join the breast-bone or sternum. Surrounding the backbone and between the ribs are muscles, so that a space, the thoracic cavity or thorax, containing the heart and lungs, is enclosed. The thorax is separated from the lower part of the body or abdomen by a muscular partition, the diaphragm, which is dome shaped.

The abdomen is the largest cavity in the body, containing the organs of digestion. It is bordered on the posterior region by the backbone and its associated muscles, and on the sides and anterior region by muscular sheets, the abdominal muscles. As the diaphragm is above the lowest rib, the abdomen has ribs in the upper part of its walls.

Below the abdomen and continuous with it is the pelvic cavity or pelvis, which is surrounded behind by the backbone and its associated muscles, and at the sides and front by the hip bones. These unite with the backbone.

The thorax, abdomen and pelvis together form the trunk. Appended to the trunk (chiefly by means of muscles) are the arms and shoulder-blades, in the thoracic region, and the legs in the pelvic region. The legs are jointed to the hip bones.

All the organs of the body are supplied with nerves and blood vessels.

B

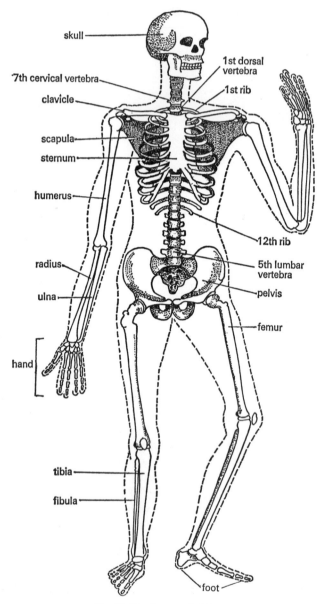

Fig. 24. The human skeleton.

An outline of the boundaries and contents of the principal body cavities is given below; in order to study them satisfactorily use must be made of diagrams and models.

Thoracic cavity

Boundaries. Behind: 12 thoracic vertebrae with their intervertebral discs.

In front: The sternum and costal cartilages.

At the sides: The ribs and the intercostal muscles.

Below: The diaphragm.

Above: The structure at the base of the neck.

Contents. The thoracic cavity contains:

The lungs, one on each side. They are covered with a membrane called the pleura, which is reflected back from their surface.

The heart between the lungs. This is covered with a membrane, the pericardium. This pericardium is reflected back and forms a sac, the pericardial sac, surrounding the heart.

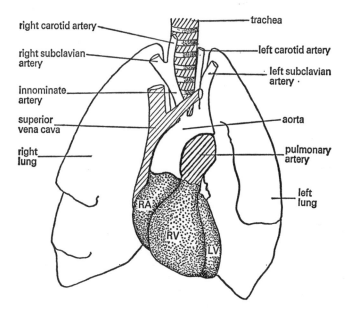

Fig. 25. The heart and the lungs (with pleura and pericardium removed). RV, right ventricle. RA, right auricle. LV, left ventricle.

The blood vessels entering and leaving the heart.   FAR-INGKS

The trachea or windpipe, which leads from the pharynx, the cavity at the back of the mouth. It divides into two at the level of the fourth thoracic vertebra, forming the two bronchi, one going to each lung.

The oesophagus or gullet, a tube leading from the pharynx to the stomach. It is situated behind the trachea and goes through the diaphragm.

The thoracic cavity also contains nerves, glands and lymphatic vessels. A large lymphatic vessel, the thoracic duct, is situated in front and slightly to the left of the vertebral column.

The pericardium, heart and large blood vessels, also the trachea, oesophagus, nerves and glands, form a partition, the mediastinum, which divides the thoracic cavity into two and separates the two lungs. The pleura as already stated is reflected back from the surface of the lungs, and covers the sides of the mediastinum partition and the corresponding half of the chest wall. In this way each lung is enclosed in its own pleural sac.

The abdomen and pelvic cavity (pelvis)

Boundaries of the abdomen. Above: The diaphragm.

Behind: The vertebral column and its associated muscles.

In front and at the sides: The abdominal muscles, the lower ribs, the iliac bones, which form part of the pelvic girdle or hip bones.

Below: The abdomen is continuous with the pelvis.

Contents of the abdomen. The abdomen contains:

The stomach, small intestine and *large intestine*, which fill most of the abdomen from the diaphragm to the pelvis (see Figs. 73 and 76).

The liver, the bulk of which lies beneath the right half of the diaphragm. It extends down and overlaps the stomach and part of the small intestine. Almost embedded in the liver is the gall bladder (see Fig. 77).

The pancreas (see Fig. 74).

The spleen (see Fig. 74).

The kidneys, which lie on the posterior wall of the abdomen. Suprarenal glands are attached to their upper ends.

Arteries, veins, lymphatic vessels, glands and *nerves*. The chief vessels are the abdominal aorta (an artery), the inferior vena cava (a vein) and part of the thoracic duct (a lymphatic vessel).

For convenience of description the abdomen is divided into nine regions by four imaginary planes, as shown in Fig. 26.

Boundaries of the pelvic cavity. Behind: Part of the vertebral column known as the sacrum and coccyx.

In front and at the sides: Part of the two hip bones and associated muscles.

Below: The floor of the pelvic cavity formed by muscles of the pelvic diaphragm.

Contents. The pelvis contains:
The urinary bladder.
The lower part of the large intestine and the rectum.
Some of the reproductive organs.

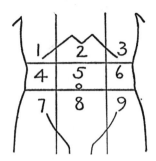

Fig. 26. Regions of the abdomen.

1, right hypochondriac. 2, epigastric. 3, left hypochrondriac. 4, right lumbar. 5, umbilical. 6, left lumbar. 7, right iliac. 8, hypogastric. 9, left iliac.

Cranial cavity

This will be described in Chapter 4.

Membranes of the body

All the cavities and tubes within the body are lined with membranes, of which there are three types:

Mucous membranes.
Serous membranes.
Synovial membranes.

Mucous membrane

A mucous membrane is found lining spaces directly communicating with the outside of the body and continuous with the skin, e.g. a mucous membrane lines the windpipe, mouth and the remainder of the gut or alimentary canal. It consists of a layer of epithelium, a layer of connective tissue and sometimes a layer of muscle. It protects underlying tissues, supports blood vessels, and sometimes absorbs and secretes.

Serous membrane

A serous membrane is found lining cavities which do not communicate with the outside of the body and surrounds the organs within these cavities, e.g. the pericardium; the pleura lining the thoracic cavity and covering the lungs; the peritoneum lining the abdominal cavity. It consists of pavement epithelium (the endothelium), and a layer of connective tissue with a network of fine elastic fibres.

The peritoneum is a double serous membrane lining the abdominal cavity (parietal peritoneum) and covering some of the viscera (visceral peritoneum). It really forms a closed sac, the peritoneal cavity, which extends from the upper part of the abdomen into the upper part of the pelvic cavity. The kidneys, which are completely outside the sac, have only

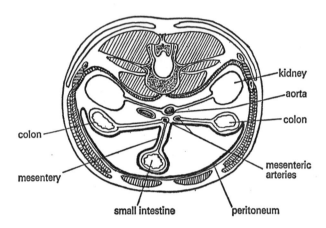

Fig. 27. Transverse section through the abdomen to show peritoneum.

their anterior surfaces covered with peritoneum. The pancreas and spleen for example, which project a little way into the sac, are incompletely covered with peritoneum, while many other organs, e.g. stomach, small intestine which project right inside the cavity, are completely covered by peritoneum which has thus become invaginated. These organs are attached to the abdominal walls by mesenteries formed of a double layer of peritoneum. Sometimes folds of peritoneum which connect various organs to the abdominal wall are called ligaments, for example the ligaments of the liver and uterus. In the mesenteries and ligaments, blood vessels, lymph vessels and nerves run to and from the organs.

Some of the organs are connected with one another by double folds of peritoneum which project into the peritoneal sac. These folds are called omenta. The two most important ones are the great omentum and the lesser omentum.

The lesser omentum separates the stomach from the lower side of the liver. The greater omentum is an apron-like fold attached to the lower border of the stomach and hanging down between the abdominal wall and the intestine.

Functions of the peritoneum. (1) It forms a smooth lining secreting a lubricating fluid and allowing without friction any necessary movement of the organs.

(2) It attaches organs together and to the abdominal wall, so keeping them in position.

(3) It stores fat, which not only forms a fat depot but helps to keep the organs warm.

(4) The great omentum wraps itself around any seat of infection in the abdomen and prevents the spread of the infection.

Synovial membrane

A synovial membrane lines cavities connected with bone. It consists of delicate vascular connective tissue covered on its free surface by an incomplete layer of epithelium. It forms a smooth slippery lining, secreting a lubricating fluid. It also absorbs foreign particles which get into joints and bits of cartilage, débris of the wear and tear of the body (see Fig. 35).

Body co-ordination

All organs and systems of the body are co-ordinated for the common good by (i) the nervous system conveying rapid impulses, which can be compared with telephone and telegraph messages, and (ii) chemical substances called hormones, carried by the blood from one organ to another. These can be compared with messages sent by post. Different hormones will be mentioned throughout the course of this book and the nervous system will be described in detail in Chapter 16, but as nervous reflexes or actions which take place involuntarily are closely connected with the functions of other organs the principle of their action will be described here.

For example, when we touch something hot with our fingers we drag our forearm away. This is not a conscious act, but takes place before we have time to think about it. Sense organs in our skin are stimulated by the heat, an impulse passes up sensory neurones to the central nervous system, across to motor neurones which convey it to the muscles, which contract and so remove the forearm from the source of the harmful stimulus. The path taken by the impulse bringing about reflex action is called a reflex arc. Thus a simple arc consists of:

(1) A sense organ or nerve ending which is stimulated.

(2) A sensory neurone along which the impulse set up by the stimulus travels to the central nervous system.

(3) A connector neurone which conveys the impulse to a motor neurone.

(4) A motor neurone which conveys the impulse to the organ which reacts.

(5) The organ which reacts.

No two neurones are structurally joined. They are separated by a minute gap filled with tissue fluid called a synapse. This will allow the impulse to go across in one direction only.

The spinal nerve consists of axons and dendrons massed together, those belonging to the sensory neurones lying dorsally or posteriorly, those belonging to the motor neurones lying ventrally or anteriorly. The spinal nerve enters the spinal cord by two roots (see Fig. 91), the sensory dendrons taking the dorsal root and the motor axons leaving by the ventral root. The nerve cells of the sensory neurones are massed together in a swelling called a ganglion on the dorsal root, the nerve cells of the motor neurone are massed together in the grey matter of the spinal cord. The white matter of the spinal cord consists of fibres cut across. These fibres convey messages to and from the brain, and thus the mind may become conscious of reflex actions which take place.

PRACTICAL

1 Study the human skeleton with the bones fixed together. Note the bony boundaries of the principal body cavities.

2 Study models of the human body.

3 Dissect a rabbit. As the structure of all mammals is similar, students should dissect a rabbit to obtain a general idea of the arrangement of the organs of the body. Details for this dissection are given in practical biology books.

Observe carefully the general arrangement of muscles and the body cavities with their contents.

[Handwritten annotations at top:]

L. LONG SHAFT —— FEMUR
I. IRREGULAR — PELVIS FACE
F. FLAT —— SCAPULA
T. TURBINATE — NOSE
S. SHORT —— WRIST

4 Skeleton and joints

[Handwritten annotations:]

S. SHAPE & SUPPORT
P. PROTECT VITAL ORGANS
A. ATTACHMENT FOR MUSCLES
R. RED CELLS PROD / MARROW (LONG SHAFT / BONES

JOINTS   BALL SOCKET   HIP
         SERRATED   SKULL
         HINGE   ELBOW

The skeleton consists of variously shaped bones held together by ligaments to form joints. These joints allow varying degrees of movement between the bones.

The bones support the body and determine its general shape; give attachment to muscles and form levers on which the muscles act; they also give protection to the more vital organs of the body, the brain, heart and lungs.

The normal skeleton is made up of about 206 bones – 34 single bones and 86 paired bones (see Fig. 24).

Each arm and shoulder	32 bones
Each leg and hip	31
Vertebral column	26
Ribs and sternum	25
Skull	22
Auditory ossicles and hyoid bone	7

Types of bones

Bones are classified according to shape as follows:

(1) *Long bones*. These bones are made up of cylinder-shaped shafts with expanded ends covered with cartilage. The shaft is hollow and is formed of compact bone; the ends are formed of cancellous bone with only a covering of compact bone.

(2) *Short bones* are cuboid, and consist of cancellous bone covered with a thin layer of compact bone.

(3) *Flat bones* are thin plates composed of three layers – two outer layers of compact bone with cancellous bone in between.

(4) *Irregular bones* have no definite shape.

The only way to acquire a knowledge of the bones is from the actual skeleton. The following short description is intended to serve as a guide to further examination. It must be remembered that a knowledge of the topography of bones is not intended merely as a test of memory. The bones are the landmarks of the body, and the position of other organs can be described in relation to the bones.

Upper extremity

The upper extremity is divided into four areas:
  (1) The shoulder
  (2) The upper arm
  (3) The forearm
  (4) The hand

The following are bones in the shoulder:

The clavicle. A long curved bone, placed in front of the root of the neck. It connects the shoulder-blade to the sternum or breast-bone. It is about 6 in long in an adult.

The scapula or shoulder-blade is a flat bone, triangular in shape, lying on the back of the ribs. It has a ridge or spine. Running along its outer part there is a thickening with a cavity called the glenoid cavity for articulation with the humerus – the bone of the upper arm.

Upper arm. There is only one bone, the humerus. It is a long bone having a rounded head which fits into the glenoid socket of the scapula. At its lower end there is a roller-shaped surface which articulates with the bones of the forearm, forming the elbow joint. The surface of the humerus is not completely smooth, as there are ridges where muscles are attached.

Forearm. There are two bones – the radius and ulna. The radius extends from the elbow to the thumb on the outer side of the wrist. The ulna extends from the inner side of the elbow to the little finger side of the wrist. The space between the two bones is filled with membrane and muscle. The projection of the elbow is formed by a projection of the ulna called the olecranon process. When the hand rotates the ulna is relatively fixed and the radius swings over it.

Hand. The bones of the hand fall naturally into three groups:
  (1) The small bones of the wrist, called carpals.
  (2) The bones of the palm, called the metacarpals.
  (3) The bones of the fingers or phalanges.

The bones of the carpus, which are about the size of large peas, are square in section and are thus short bones. The metacarpals and phalanges, although only an inch or so in length, belong to the class of long bones.

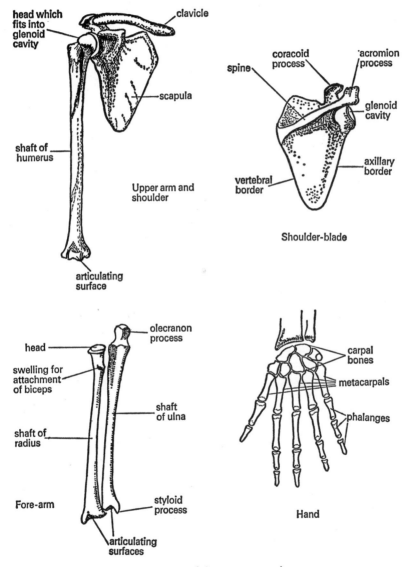

Fig. 28. Bones of the upper extremity.

Lower extremity

The leg or lower extremity is divided into the following areas:
    (1) Hip region.
    (2) Thigh region.
    (3) Leg and foot.

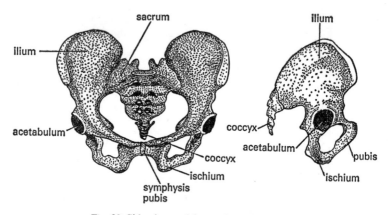

Fig. 29. Side view and front view of the pelvis.

The hip bones are two large, irregular-shaped bones, each made up of three parts, the ilium, ischium and pubis, which meet at the acetabulum, a cup-shaped cavity on the outer side. The two hip bones meet in front at the symphysis pubis and posteriorly they join the sacrum. This bony ring or girdle forms the pelvis, which transmits the weight of the body to the limbs and protects such organs as the rectum and the urinary bladder, and female reproductive organs. The acetabulum receives the head of the femur to form the hip joint.

The femur or thigh bone extends from the hip to the knee and is the largest bone in the body. Its upper end is set at an angle to form a neck, the head of which fits into the acetabulum, forming the hip joint. From the neck of the femur the shaft leads to the expanded lower end, known as the condyles. With the knee-cap and the upper end of the tibia the condyles form the knee joint.

The knee-cap or patella is a small bone formed in the tendon, which protects the knee joint.

The leg proper extends from the knee to the ankle and contains two bones; a strong thick bone, the tibia, lies on the inner or big toe side of the leg. It consists of a shaft, and expanded ends. The upper end has two condyles which articulate with the femur to form the knee joint. The expanded lower end forms, with the other bone of the leg, the upper bones of the ankle joint.

The second bone of the leg, the fibula, lies on the outer side of the leg, alongside the tibia. It is a more slender bone than the tibia. Its upper end takes no part in the formation of the knee joint. Its lower end forms the outer part of the ankle joint.

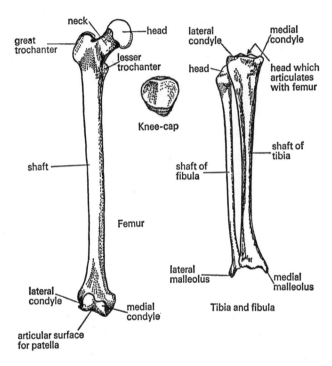

Fig. 30. Bones of the thigh and leg proper.

Foot. The general structure of the foot is similar to that of the hand. It can be divided into three areas. The tarsus or hinder part of the foot consists of seven short bones forming the ankle region. The middle part of the foot, corresponding to the arch of the foot, is made up of five short 'long' bones termed the metatarsals.

The bones of the toes are also termed phalanges – three for the four outer toes, and two for the great toe.

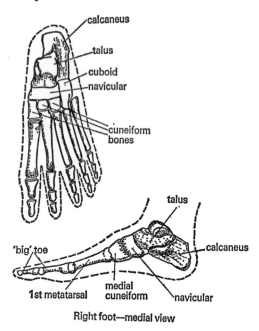

Right foot—medial view

Fig. 31. The foot.

Vertebral column

The vertebral or spinal column forms the 'backbone'. It is made up of
thirty-three small bones called vertebrae. The first twenty-four are
separate and movable but the last nine are fused into two groups.

The vertebral column is divided into five regions:

Cervical or neck region	7 vertebrae	
Dorsal or thoracic region	12	⎫ movable
Lumbar region	5	⎭
Sacrum SA- KRUM	5	⎫ fused.
TAIL Coccyx KOK-SIKS	4	⎭

and shows four curves:

(1) In the cervical region, convex forward.
(2) In the thoracic region, concave forward.
(3) In the lumbar region, convex forward.
(4) In the sacral and coccygeal regions, concave forward.

The vertebrae of the sacrum and coccyx are fused to form single bones. They form the posterior part of the pelvis.

The size and shape of the single vertebrae vary. The cervical vertebrae are thin and light bones, while the lumbar bones are thick and heavy. Discs of cartilage are placed between the bones of the vertebral column, making the whole 'backbone' elastic and allowing bending movement.

A typical vertebra consists of:

(1) A strong mass of bone termed the body or centrum.
(2) A bony arch.
(3) Several bony projections.

When the vertebrae are placed upon each other the arches form a tube, the neural canal, which contains the spinal cord, a delicate organ, which is thus protected by the bony arches and bodies of the vertebrae.

The sacrum is a single wedge-shaped bone formed by the fusion of five sacral vertebrae. The coccyx – which is the rudimentary tail found in human beings – is formed by the fusion of four small vertebrae.

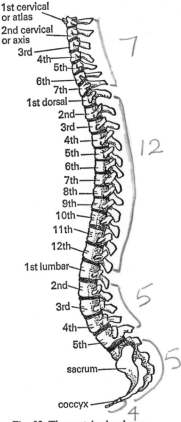

Fig. 32. The vertebral column.

Bones of the thoracic cavity

The bones forming the framework of the cage-like thoracic cavity which contains the heart and lungs are:

(1) The dorsal or thoracic vertebrae.
(2) The ribs. There are twenty-four ribs, twelve on each side. Each rib articulates with usually two dorsal vertebrae behind, and, excepting the last five pairs, with the sternum or breast-bone by their costal cartilages in front. The lower two ribs, which end freely in front, are termed 'floating ribs'. The eighth, ninth and tenth pairs articulate with the rib above

them. The ribs are curved long bones; each pair forms a circular arch called the costal arch.

(3) The sternum. This is a long flat bone said to be shaped like a dagger. The lower part is composed mainly of cartilage, making the sternum flexible.

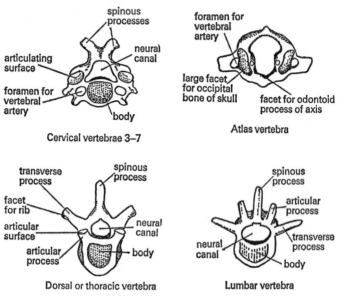

Fig. 33. Human vertebrae (the atlas is the first cervical vertebra).

Skull

The skull may be divided into two portions:
  (1) The cranium, containing the brain.
  (2) The face, with its air sinuses and cavities.

The cranium is a large bony case which protects the most delicate organ of the body, the brain. The walls of the 'box' are perforated at many points by foramina, to allow vessels and nerves to pass to and from the brain.

The cranium consists of the following bones:

(1) *The occipital bone,* which forms the back part of the base of the skull. There is a large opening, the foramen magnum, through which the brain connects with the spinal cord.

The occipital bone articulates with the uppermost vertebra, in front with the temporal and sphenoid bones, and above with the parietal bones.

(2) *The sphenoid* is the central bone forming the base of the skull. It has wing-like extensions passing outwards which support a great part of the base of the brain.

This bone is pierced by many channels or foramina through which pass the vessels and nerves. It helps to form part of the eye socket.

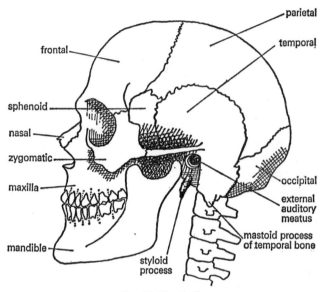

Fig. 34. The skull.

(3) *The frontal bone* is fused with the front part of the sphenoid. It is made up of a horizontal portion forming the floor of the cranium and the roof of the orbit, and a rounded ascending portion forms the bone of the forehead.

(4) *The ethmoid* is a small bone, placed between the sphenoid and frontal bones. It forms the roof of the cavity of the nose and helps to complete the floor of the cranium.

(5) *The temporal bones* lie on the sides of the cranium, and are located between the occipital bone and the sphenoid. They do not meet in the central part of the base of the skull. The temporal bones contain the bony parts of the external, middle and inner ear, and some air cells, the mastoid cells, which communicate with the middle ear.

(6) *The parietal bones* form the roof and sides of the cranium.

The thickness of the various skull bones varies. Where the bony wall is thin the bone is so placed that it is protected from injury by its deep position. The exposed bones are thick and strong. The brain is thus assured of effective protection.

The face. The principal bones, which give shape to the face, are:

(1) *The frontal bone* – forming the brow and upper part of the orbital cavity.

(2) *The temporal bone*, which articulates with the *zygomatic bone* which forms the prominence of the cheek.

(3) *The nasal bone* – forming the bridge of the nose.

(4) *The upper jaw*, which carries the teeth and is formed by the irregular maxillary bones, which articulate in the mid-line forming the roof of the mouth.

(5) *The lower jaw or mandible*. The mandible is made up of two halves which fuse in the mid-line to form the prominence of the chin. Each half is composed of a horizontal part which carries the lower teeth, and a vertical part which articulates with the temporal bone immediately in front of the ear.

The lower jaw is mobile; the movement of this bone assists in mastication, speaking, etc.

Some of the bones of the skull such as the frontal bone, the maxilla, and ethmoid bones have cavities in their interior known as air sinuses. These sinuses have the effect of making the bones lighter than they otherwise would have been. They communicate with the nasal cavities and are lined with mucous membrane, which is continuous with the mucous membrane lining the nasal cavities. They may be infected by the direct spread of organisms from the nose. Infection may reach the brain from the sinuses lying near the cranium.

Articular system or joints

Joints are formed whenever bones come into direct contact with each other. They can be divided into three classes:

Immovable joints. In this type of joint no movement takes place between the bones. In a fibrous immovable joint the bones are connected by connective tissue which blends with the periosteum, for example, a joint known as suture between two flat bones of the skull. In a primary cartilaginous immovable joint, which is usually temporary in nature, the bones are connected by cartilage, for example, the joint between the shaft and epiphysis of a long bone in a young person.

Slightly movable joints (*secondary cartilaginous joints*). In this type of joint the opposed bony surfaces are covered by hyaline cartilage and connected by a fibro-cartilaginous pad. Two examples are the joints between the bodies of adjacent vertebrae, and the joint between the two pubic bones of the pelvis. The elasticity of the intervening cartilage of these joints allows slight movement. In the female pelvis this is important during childbirth, where a slight 'give' allows the head of the child to pass through.

Movable joints. The general structure of such a joint can best be illustrated diagrammatically. It is made up of the following tissues:

(1) *Cartilage*, known as articular cartilage, covering the surface of the part of the bone which enters into the joint. In health, true bone does not enter into the actual make-up of the joint.

(2) *The capsule*, which consists of strong fibrous tissue, attached to the rim of the articular cartilage. The capsule encloses the 'joint cavity'. Parts of the capsule may be thickened to form definite bands or ligaments; these ligaments are developed to give the joint support in special directions.

(3) *The synovial membrane.* This membrane, made up of epithelial cells, lines the inside of the joint cavity. It secretes an oily, milky fluid, facilitating smooth movement.

(4) *Blood vessels, lymph vessels and nerves* supplying the joints.

The type of movement at a joint may also be used to classify joints:

(1) *The ball and socket joint* allows one bone to move freely in all directions. The best examples of this are the shoulder joint and the hip joint. The rounded heads of the movable bone – humerus and femur – form a ball which fits into the socket found on the scapula and the pelvis respectively.

There are two opposing properties of a ball and socket joint, (*a*) stability, (*b*) mobility. The shoulder joint has extreme mobility, having a shallow cup in the scapula. The hip joint, with a deep cup in the pelvis, has not such free movement, but is much more stable. Dislocations of the shoulder are common, those of the hip joint very rare.

(2) *The gliding joint* allows a small amount of movement only, two flat surfaces moving on one another. Examples of gliding joints are the joints between articular processes of two vertebrae.

(3) *The hinge joint* allows movement only in one plane. Examples of

(HINGE JOINT)

this are seen in the elbow and the joint between two phalanges of the fingers, where a convex cylindrical surface meets a concave cylindrical surface. The knee joint functions as a hinge joint though it has the structure of a condyloid joint.

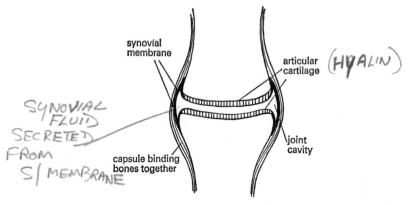

Fig. 35. Section of a simple joint.

(4) *The condyloid joint* allows movement in two planes. Examples are the joint between the skull and the lower jaw and the joint between the skull and the first cervical or atlas vertebra.

(5) *The pivot joint* allows rotation. Examples are the joint between the first (atlas) and the second (axis) cervical vertebrae and the joints between the radius and ulna which allow the radius to swing over the ulna.

Leverage

A lever is a strong bar which moves about a fixed point known as the fulcrum and transmits a force, thus allowing the force (effort) to be applied indirectly to an object (load). Within the body many bones, especially long bones, act as levers and the muscles supply the effort.

There are three classes of levers:

(1) *Class 1* in which the fulcrum is situated between the effort and the object, e.g. a seesaw. Levers of this class work within the body to give stability. For example, the atlas acts as a fulcrum about which the head turns, the load being the front or the back of the head. The effort is supplied by the muscles of the neck.

(2) *Class 2* in which the load is situated between the fulcrum and the effort, e.g. a wheelbarrow. Few examples of this type of lever are found

in the body, but one is seen in the foot during walking. The fulcrum is formed by the heads of the metatarsals, the load is the body weight and the effort is supplied by the calf muscles attached to the heel bones.

(3) *Class 3* in which the effort is applied between the fulcrum and the load, e.g. coal tongs (double lever). Although there is never mechanical advantage in using this type of lever there is convenience in handling.

Most levers in the body belong to this class; for example, the forearm moving about the elbow joint which forms the fulcrum. The load is any object in the hand or the weight of the hand itself and the effort is provided by the biceps muscle which acts on the upper end of the forearm.

PRACTICAL

1 Examine a complete articulated skeleton. Note the positions of the different types of joints.
2 Examine the shoulder and arm. Draw careful diagrams.
3 Examine the pelvis and leg. Draw careful diagrams.
4 Examine the backbone. Draw careful diagrams of:
  (a) the atlas vertebra (the first cervical);
  (b) the axis vertebra (the second cervical);
  (c) one of the remaining cervical vertebrae;
  (d) a thoracic vertebra;
  (e) a lumbar vertebra;
  (f) the sacrum and coccyx.
5 Examine and make diagrams of the skull.

MOVEMENTS OF THE HIP. JOINT

ABDUCTION (AWAY)

ADDUCTION (TOWARDS)

FLEXTION (BENDING)

EXTENSION (STRAIGHTENING)

5 Voluntary muscular system

The structure of voluntary muscles has been described in another chapter. Collectively they form the muscular system, whose function is to bring about co-ordinated movement. Muscles form the flesh of the body.

Muscles require oxygen and food to produce energy for the work they perform. They therefore have a liberal blood supply which is adjusted according to the immediate need. The co-ordination of movements is brought about by nerves, and each muscle has its own nerve supply. The nerves include both motor fibres, which activate the muscles, and sensory fibres, which indicate the position of the muscle, thus giving the individual muscle sense. Unlike involuntary muscles and the heart, a voluntary muscle will contract in the intact body only if it is connected to the central nervous system by its nerve. It is never completely at rest but in a state of tone or tension ready to respond quickly to a stimulus. The reason is that some nerve impulses are always passing to it so that at any one time a few fibres are contracted. Usually in the body when a muscle is stimulated to contract, successive impulses pass to the muscle so that there is not a single twitch but a sustained contraction, known as a tetanus. When a muscle undergoes prolonged stimulation it becomes fatigued, its contractions getting slower and less strong. There is always a very short period, known as the latent period, between the time the stimulus is received by a muscle and the contraction begins. When muscular fatigue develops the latent period becomes longer.

Muscles bring about their movement by means of their attachment to bones or cartilage which act as levers. Muscles of special function are attached to other structures, e.g. muscles of expression are attached to the skin of the face, the ocular muscles are attached to the eyeball. Sphincter muscles, which are composed of muscle fibres running circularly round an opening and on contraction close the opening, are in some cases composed of striated fibres and under control of the will, e.g. external anal sphincter, which closes the anal aperture.

Muscle tissue itself is not suitable for attachment to bones, as it is liable to tear. Connective tissue, specialized into strong tendons and merging into the muscles, gives the muscles strong attachments. Sometimes muscles are attached by more flattened tendon-like structures called aponeuroses. Membranes of connective tissue called deep fasciae (the superficial fascia lies under the skin and connects the skin to the underlying parts) may form sheaths for muscles and bind muscles together. In some places a deep fascia is thickened to form a strong protective band, for example, the annular ligaments of the wrist (see page 59).

Most muscles have at least two attachments. One attachment, known as the origin, is usually the more stable, the other, known as the insertion, the more mobile. These terms, origin and insertion, can only be relative, as muscles may at times function in the opposite direction. For example, the breast or pectoral muscles normally draw the arm to the side, the chest wall being relatively fixed. In forced respiration, however, a person may fix his arms by gripping a support and the pectoral muscles then assist the muscles of respiration. It is common knowledge that a person gasping for breath always tries to find something to grip.

The muscles in the skeleton do not act individually, but in groups. This enables fine movement to be performed. Certain muscles, which may not take part in the production of an actual movement, assist such a movement by fixing a bone or bones in a suitable position; for example, when the arm is raised from the side, muscles of the shoulder joint contract to hold the scapula and other bones steady. Such muscles are known as fixation muscles. Others, known as synergists, steady joints during movement; for example, when the flexors of the fingers contract, the extensors of the wrist give rigidity to the wrist joint.

A movement does not only involve the contraction of one group of muscles. The muscles known as antagonists which oppose such a movement must relax; such co-ordination is one of the functions of the nerve supplying the muscle. Sometimes opposing groups of muscles are contracted equally and no movement takes place, the contracting muscles having a supporting effect. This effect is important in maintaining posture.

In the short descriptions of the principal muscles various terms are used to describe the action.

Neck and trunk muscles

(1) *Flexion and extension.* Flexion indicates that the angle between the axes of the bones is decreased, and extension that the bones are brought into more or less a straight line, e.g. bending and straightening of the back.

(2) *Rotation.* The bones are rotated one over another, e.g. turning of the trunk.

Arm muscles

For descriptive purposes the arm is considered as being at the side of the body, palms to the front and thumb pointing outwards.

(1) *Flexion and extension* are used in the same sense as in the trunk.

(2) *Adduction* means the arm is moved towards the middle line of the body, *abduction* that it is moved away from the middle line. *Circumduction* means the whole limb is moved in a circular manner.

(3) *Internal rotation* means the front of the arm is rotated towards the middle of the body; *external rotation* that the front of the arm is rotated away from the middle line of the body.

(4) *Pronation* indicates the palm of the hand is rotated to face backwards; *supination* that the palm is rotated to its original position and faces forwards.

Muscles of the leg

The same terms are used as in the description of arm movements, with the exception of pronation and supination. Movements of the foot are not so wide in range as are those of the hand. The foot is said to be:

(1) *Inverted* when the sole is turned inwards towards the mid-line of the body.

(2) *Everted* when the sole is turned in the opposite direction.
The movements of the foot on the leg are described as:

(1) *Dorsi flexion* when the foot is cocked up.

(2) *Plantar flexion* when the toe is pointed as in toe dancing.

The importance of being familiar with the origin and insertion of a muscle is that it explains the function of a muscle. The best way to study muscles is, of course, in the dissecting-room. When this is not possible the position of muscles should be studied on the skeleton with the aid of diagrams. The student will find much interest in performing the various movements noting the muscles which are brought into action. These muscles become tense, and if the movement is continued a feeling of stiffness may develop.

A short description of the various muscles is found in the remainder of this chapter, but a book cannot do more than give a general idea of their arrangement.

Muscles of the face and head

The function of these muscles is to give facial expression and to bring about mastication of the food. It is not possible to make a hard and fast distinction, as many muscles take part in both functions.

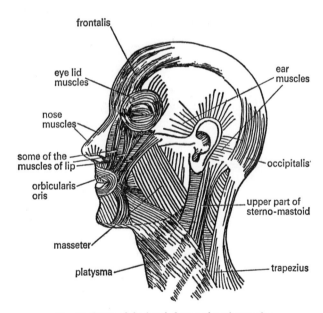

Fig. 36. Some of the head, face and neck muscles
(the frontalis and occipitalis together form the occipito-frontalis muscle).

Name	Position and attachments	Action	Cranial nerve supply
Temporal	Covers temporal region of skull	Raises lower jaw, brings about mastication	5th
Masseter	Side of face	Brings about mastication	5th
Pterygoid	do.		

Name	Position and attachments	Action	Cranial nerve supply
Buccinator	Forms cheeks	Brings about mastication and whistling	7th
Tongue muscles	Tongue	Move tongue and assist mastication	9th
Occipito-frontalis	Over front of brow	Wrinkles forehead, gives expression	7th
Orbicularis oris	Surrounds mouth	Closes mouth, gives expression	7th
Orbicularis oculis	Surrounds eye	Move eyelids; expression	7th
Muscles of nose and lips	Surround nose and lips	Dilate nose; open lips; give expression	7th

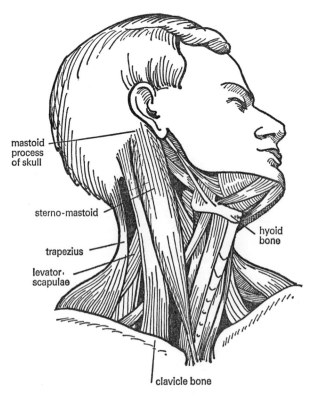

Fig. 37. Muscles of the right side of neck.

Muscles of the neck

Their main function is to move the head. In forced respiration they assist
the respiratory muscles.

Name	Position and attachments	Action	Nerve supply
Sterno-mastoid	At each side of neck. Arises from sternum and clavicle and is attached to mastoid process in skull	Moves head to side. If working together bend head on chest	11th cranial
Platysma	Beneath skin of neck and attached to it. It reaches to corners of lips	Moves skin of neck; draws down corners of mouth; gives expression	7th cranial

Muscles of the chest

The muscles attached to the chest cause the respiratory movements and
move the head, neck and arms. As already mentioned, any muscle
attached to the chest may assist respiratory movements during forced
respiration.

Name	Position and attachments	Action	Nerve supply
Pectoralis major	Front of chest. Arises from ribs, sternum and clavicle. Inserted into shaft of humerus	Adducts arm to side and carries it across the chest	From brachial plexus
Pectoralis minor	Upper part of chest, deeper than pectoralis major. Attached to ribs and inserted on coracoid process of scapula	Pulls scapula forward	do.
Serratus anterior	On side of chest. Arises from ribs. Inserted in scapula	Pulls scapula forward	do.
External and internal intercostal muscles	In between the ribs; fibres of two muscles cross	Bring about respiratory movements	Intercostal nerves

The diaphragm is one of the chest muscles and requires special description. It is a dome-shaped muscle separating the thoracic from the
abdominal cavities. It arises from:
(1) the lumbar vertebrae;
(2) the back of the lower end of the sternum;
(3) the inner side of the lower six pairs of ribs and it is inserted on a
central tendon.

The diaphragm is the chief respiratory muscle. (See page 102 for an account of the mechanism of respiration.)

Respiration is the main function of the diaphragm, but it also increases the abdominal pressure, assisting in micturition, defaecation, and childbirth.

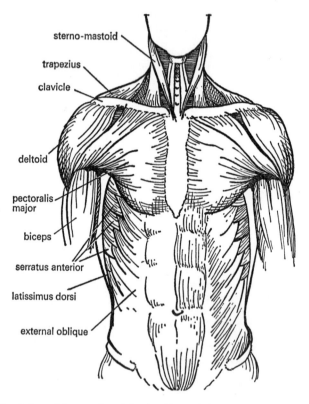

Fig. 38. Front view of the muscles of the shoulder and chest and the superficial muscles of the abdomen.

The heart and lungs lie above the diaphragm, and the liver, stomach, spleen, suprarenal glands and kidneys immediately below it. The muscle is pierced by (1) the aorta, (2) the inferior vena cava, and (3) the oesophagus.

The muscle is supplied by the phrenic nerve from the cervical plexus.

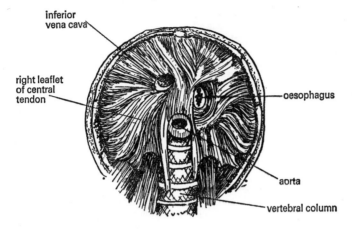

Fig. 39. The under-surface of the diaphragm.

Abdominal muscles

The abdominal muscles, which lie on the front of the abdomen, have two functions:

(1) They move the trunk and assist in acts of respiration, defaecation, etc.

(2) They protect the delicate abdominal organs.

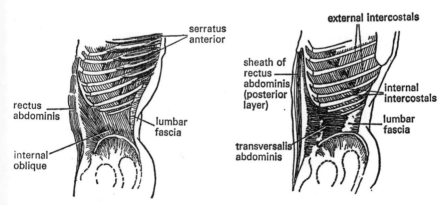

Fig. 40. The middle layer of abdominal muscles (side view).

Fig. 41. The deep layer of abdominal muscles (side view).

Name	Position and attachments	Action	Nerve supply
Rectus abdominis	A flat muscle in the front of the abdomen. Attached to pubis and passes upwards to be inserted on 5th, 6th and 7th ribs. It is separated from its fellows of the opposite side by the linea alba[1]	Flexes trunk	Intercostal
External oblique	A flat muscle in front of the abdomen. It arises from the lower ribs and passes downwards and forwards to be inserted into the sheath, enclosing the rectus muscle, and the iliac crest	Flexes trunk laterally; rotates trunk	Intercostal
Internal oblique	A flat muscle arising mainly from the iliac crest and running upwards and forwards crossing the external oblique muscle. Inserted on the sheath of the rectus and lower ribs	Flexes trunk laterally; rotates trunk	Intercostal
Transversalis abdominis	A flat muscle which runs transversely deep to the oblique muscles	Assists oblique muscles	Intercostal
Quadratus lumborum	Lies on side of vertebral column passing from iliac crest to lowest rib	Extends vertebral column	From lumbar plexus

The aponeurosis of the external oblique muscle is a thin strong membrane which medially ends in the linea alba. It is continuous with its fellow of the opposite side. Below it forms a thick band known as Poupart's or the inguinal ligament, which runs obliquely downwards and forwards from the ilium to the pubis. A canal runs through the muscles just above Poupart's ligament, carrying, in the male, the spermatic cord, together with blood vessels and nerves. Abdominal contents may pass down the canal and cause a swelling on the abdominal wall. The condition is known as a hernia or rupture.

Muscles of the floor of the pelvic cavity

The pelvic floor is formed on either side by the levator ani and coccygeus muscles. The levator ani is a broad flat muscle which arises from the inner surface of the pelvic cavity and passes downwards to unite in the mid-line with its fellow of the opposite side. The coccygeus lies behind the levator ani arising on the ischial part of the pelvic cavity and passing to the sacrum and coccyx.

The pelvic floor forms a muscular diaphragm which supports the pelvic

[1] The linea alba is a white tendinous line in front of the abdomen stretching from the sternum to the pubis.

organs – the bladder and rectum and, in the female, the uterus and vagina. The levator ani muscle also helps in defaecation.

When the abdominal pressure goes up as a result of the contraction of the abdominal muscles during muscular exertion, the pelvic diaphragm contracts to prevent the abdominal contents being pushed into the pelvic cavity. During childbirth, when the abdominal muscles contract strongly the pelvic diaphragm relaxes.

Muscles of the back

The functions of the back muscles are:
(1) To support the spine.

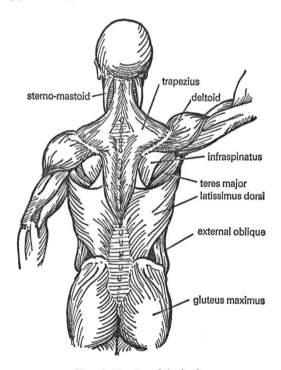

Fig. 42. Muscles of the back.

(2) To move the spine in all directions.
(3) To move the extremities of the body, the arms, limbs, head and neck.

MUSCLES OF THE BACK

Name	Position and attachments	Action	Nerve supply
Trapezius	A broad, flat muscle situated at back of neck and shoulders. Arises from occipital bone and vertebrae. Inserted into clavicle and spine of scapula	Pulls head back; braces shoulder	Spinal accessory and nerves from brachial plexus
Latissimus dorsi	Flat muscles over lower part of chest and loins. Arises from lower dorsal vertebrae and crest of ilium. Inserted by a tendon into front of an upper part of humerus	Adducts and pulls humerus back. Rotates arm inwards If arm is fixed, raises body	From brachial plexus
Erector spinae	Deep thick muscles on each side of spine	Straightens spine	Posterior rami of dorsal nerves

Muscles of the shoulder and arm

The muscles of the shoulder and upper arm are concerned with the coarse arm movements; the forearm and hand muscles are specialized for fine movements.

MUSCLES OF THE SHOULDER AND UPPER ARM

Name	Position and attachments	Action	Nerve supply
Deltoid	Forms point of shoulder. Crosses from outer part of clavicle and spine of scapula. Inserted in outer side of shaft of humerous	Adducts arm	From brachial plexus
Subscapularis	Lies between ribs and scapula from which it arises. Inserted into humerus	Internal rotator of humerus	do.
Supraspinatus	Lies above spine of scapula and is inserted into humerus	Assists deltoid	do.
Infraspinatus	Lies below spine of scapula. Inserted into humerus	External rotator of humerus	do.

Name	Position and attachments	Action	Nerve supply
Teres major and minor	Runs from lower end of scapula to humerus	Draw arm down and backwards	From brachial plexus
Biceps	A long muscle which has two origins from the scapula. Inserted into upper end of radius	Flexes elbow	do.
Coraco-brachialis	Arises from scapula and inserted into humerus	Adducts arm	do.
Brachialis anterior	Arises from front of humerus. Inserted into ulna	Flexes elbow	
Triceps	Has three origins, one from scapula and two from humerus. Inserted into olecranon process of ulna	Extends elbow	do.

MUSCLES OF FOREARM AND HAND WHICH FLEX THE ELBOW OR EXTEND THE WRIST OR FINGERS AND SUPINATE FOREARM

Name	Position and attachments	Action	Nerve supply
Brachio radialis	Arises from humerus; inserted into radius	Flexes elbow; supinates forearm	From brachial plexus
Extensor carpi radialis longus	Arises from a point at lower end of humerus. Inserted into 2nd meta-carpal	Extends wrist	do.
Extensor carpi radialis brevis	As above to 3rd meta-carpal	do.	do.
Extensor digitorum communis	As above to 2nd and 3rd phalanges of fingers	Extends wrist and fingers	do.
Extensor digiti minimi	As above and extends to 1st phalanx of little finger	Extends wrist and little finger	do.
Extensor carpi ulnaris	As above and extends to base of 5th metacarpal	Extends wrist	do.
Supinator	Twists round upper part of radius to front of radial shaft	Supinates fore-arm	

The thumb has special extensor muscles.

The deep extensors of the thumb pass from the back of the radius and ulna to the carpo-metacarpal joint, the metacarpal-phalangeal joint and the inter-phalangeal joint of the thumb.

c

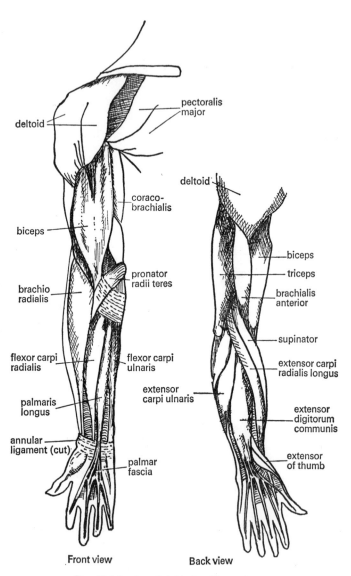

deltoid

pectoralis major

coraco-brachialis

biceps

pronator radii teres

brachio radialis

flexor carpi radialis

flexor carpi ulnaris

palmaris longus

extensor carpi ulnaris

annular ligament (cut)

palmar fascia

deltoid

biceps

triceps

brachialis anterior

supinator

extensor carpi radialis longus

extensor digitorum communis

extensor of thumb

Front view

Back view

Fig. 43. Muscles of right shoulder and arm.

MUSCLES WHICH FLEX THE WRIST
AND FINGERS AND PRONATE THE FOREARM

Name	Position and attachments	Action	Nerve supply
Pronator radii teres	From internal side of lower end of humerus to outer surface of shaft of radius	Pronates hand	From brachial plexus
Flexor carpi radialis	From humerus as above to front of hand	Flexes wrist	do.
Palmaris longus	do.	do.	do.
Flexor carpi-ulnaris	do.	do.	do.
Flexor digitorum sublimis	From humerus as above to 2nd phalanx of finger	Flexes wrist and finger	do.
Flexor digitorum profundus	From front of shaft of ulna to terminal phalanx of fingers	do.	do.
Flexor pollicis longus	From front of shaft of radius to terminal phalanx of thumb	Flexes wrist and thumb	do.
Pronator quadratus	A flat muscle across hand end of radius and ulna	Very weakly pronates hand	do.

The wrist and hand are complicated structures. In addition to bones and the muscle tendons, the following structures are present:

(1) Palmar fascia – deep fascia binding the deep structures together.

(2) Annular ligaments – thick bands of deep fascia surrounding the wrist and holding the extensor and flexor tendons in position.

(3) Synovial sheaths which surround the muscle tendons and lubricate their movement.

(4) Muscles at the base of the thumb forming the thenar eminence, and muscles at the base of the little finger forming the hypothenar eminence.

(5) Lumbricales muscles deep in the palm, and the interossei muscles between the metacarpal bones. These muscles are concerned with fine movement, such as writing.

Two anatomical spaces are described in the arm:

(1) The axilla or armpit, lying between the chest wall and the humerus, and the muscles attached to these structures.

(2) The cubital fossa at the bend of the elbow.

Muscles of the hip and leg

The function of the leg muscles is to bring about the various movements of walking, etc. Those of the hips and thigh are well developed and give strength and stability to the region. The foot and leg do not have to carry out the fine movements required of the forearm and hand, and the small muscles in the region are not so highly developed as in the hand.

MUSCLES OF HIP AND THIGH

Name	Position and attachments	Action	Nerve supply
Ilio-psoas	Lies at back of abdominal cavity. Arises from lumbar vertebrae and ilium. Inserted into upper end of femur. Tendon passes under Poupart's ligament	Flexes thigh	Femoral nerve from lumbar plexus
Quadriceps femoris, consisting of: (1) rectus femoris (2) vastus internus (3) vastus externus (4) vastus intermedius	Front of thigh arising from ilium and shaft of femur. Inserted by a common tendon (the patella is in this tendon) into tibia	Extends knee	do.
Sartorius	Long narrow muscle. Arises from ilium and passes obliquely across thigh. Inserted into tibia	Assists in action of crossing legs	do.
Adductors: (1) pectineus (2) adductor longus (3) adductor magnus (4) adductor brevis	Arise from pubis and run down inner side of thigh to shaft of femur	Adduct and externally rotate thigh	Obturator nerve from lumbar plexus
Hamstrings, biceps, semi-membranosus, semitendinosus	On back of thigh, arise from ischium and are inserted into tibia and fibula by tendons	Extend thigh and flex knee	Great sciatic nerve from sacral plexus

Name	Position and attachments	Action	Nerve supply
Buttock or gluteal muscles: (1) gluteus maximus (2) gluteus medius (3) gluteus minimus	Form prominence of buttocks. Arise from sacrum and ilium and are inserted into trochanters and shaft of femur	Raise trunk from stooping. Adduct thigh	Gluteal nerves from sacral plexus

MUSCLES OF LEG

Name	Position and attachments	Action	Nerve supply
Tibialis anterior	Front of leg on inner side of tibia. Inserted into tarsal bones	Dorsi-flexes foot and inverts foot	Branch of lateral popliteal nerve
Extensor hallicus longus	Runs from fibula to great toe	Dorsi-flexes foot. Extends great toe	do.
Extensor digitorum longus	From fibula to all toes except great toe	Dorsi-flexes foot. Extends toes	
Peroneal muscles: p. longus, p. brevis, p. tetus	From fibula to sole of foot	Evert foot	Branch of lateral popliteal nerve
Gastro-cnemius	The calf muscle. Arises from condyles on lower end of femur and inserted by tendo Achillis on to the long tarsal bone which forms the heel	Plantar flexes the foot and throws body forward when walking	Medial popliteal
Soleus	Lies deep to gastro-cnemius and its tendon joins tendo Achillis	Plantar flexes the foot and throws body forward when walking	Posterior tibial
Popliteus	Lies behind the knee, attached to femur and tibia	Flexes knee	Medial popliteal
Flexor hallucis longus	At back of fibula to big toe	Plantar flexes foot. Flexes great toe	Posterior tibial
Flexor digitorum longus	Back of fibula to 4 smaller toes	Plantar flexes ankle and flexes toes	do.

Name	Position and attachments	Action	Nerve supply
Tibialis posterior	Lies deep at back of leg and passes to lower surface tarsal bones	Inverts foot	Posterior tibial

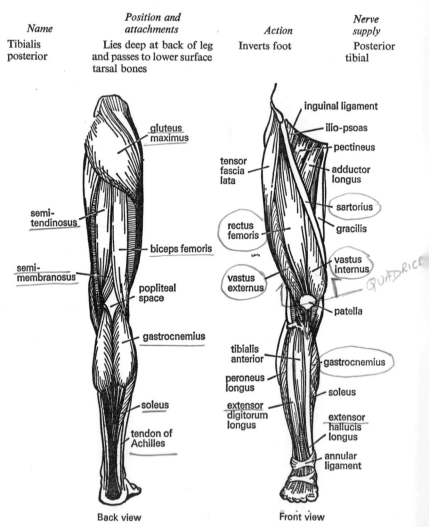

Back view   Front view

Fig. 44. Muscles of the leg.

The foot has a complicated structure similar to the hand. As there is no necessity for fine movements in the foot the small muscles are not developed as in the hand.

The leg has two anatomical spaces:

(1) Scarpa's triangle, situated just below the inguinal or Poupart's ligament. It is analogous to the axilla.

(2) Popliteal space at the back of the knee. It is analogous to the cubital fossa.

PRACTICAL

1 Determine with the aid of diagrams the position of muscles on the skeleton.

2 Perform various movements and note which muscles are brought into action.

6 The blood

General scheme of circulation

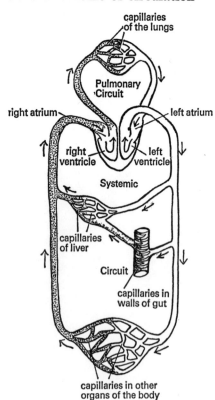

Fig. 45. Diagram to illustrate the scheme of circulation.

Blood is a reddish fluid tissue which circulates round the body, supplying food and oxygen to the other tissues and removing waste products. It circulates in closed vessels and is kept moving by a muscular pump, the heart.

Arteries are vessels through which blood passes away from the heart. They give off smaller branches, and a small artery goes to every single organ in the body. When it reaches an organ it divides up into smaller branches called arterioles, and finally these break up into a network of small thin-walled tubes called capillaries. These ramify through the organ and join up to form small veins or venules. These venules join until finally a small vein takes the blood away from the organ. It is while the blood is in the thin-walled capillaries that the interchange between the blood and the tissue takes place. The small veins join up to form large veins through which blood returns to the heart.

The diagram of Fig. 45 illustrates the scheme of circulation. It shows how the heart is divided into four chambers and that there are two circuits, the pulmonary and systemic, the blood passing through the heart twice on its way round the body.

Structure of the blood

When examined under the microscope the blood is seen to consist of a liquid with solid bodies in it. The liquid is called plasma, and the solid bodies are blood corpuscles and blood platelets.

Plasma

The plasma is a yellow, slightly alkaline fluid, having a specific gravity of 1·028 (water 1·000). It consists of about 90 per cent water and 10 per cent solid matter in solution. The total concentration of the dissolved solid matter remains practically constant, so that the osmotic pressure of the blood plasma only varies within a narrow limit. The composition of the blood plasma is controlled chiefly by the kidney.

The solid matter includes:

Proteins – serum albumin, serum globulin, fibrinogen, prothrombin. These do not vary much in amount.

Amino acids. The amount varies according to the food and the lapse of time from the intake of proteins.

Fats. The amount of fat in the blood is high after a fatty meal and low during starvation.

Glucose. The amount remains practically constant at about 0·1 per cent (0·175 per cent – 0·075 per cent). If it varies beyond these limits a serious condition is produced (see Chapter 13).

Urea and other nitrogenous waste. The amount varies with the amount of protein in the diet and the rate of destruction of living matter. Soon after it passes into the blood it is excreted by the kidneys, and in health the concentration of these substances in the blood is never allowed to rise high.

Salts, including sodium chloride, magnesium sulphate, calcium chloride, sodium bicarbonate, and phosphates. The concentration of the individual salts remains practically constant.

Corpuscles

There are two kinds of corpuscles, red corpuscles and white corpuscles. The red corpuscles, numbering about 5,000,000 per mm^3 of blood (in the whole body there are 25 trillion), give the red colour to the blood, although when examined singly they are pale buff. In shape they are like biconcave discs, and in abnormal blood they tend to collect to form piles of red cells or 'rouleaux'. This is known as agglutination.

The red blood cells have no nucleus. An envelope encloses the colouring matter haemoglobin; haemoglobin is a complex protein, purply red in colour and contains a large amount of iron. Its chief property is to combine with oxygen to form oxyhaemoglobin, which has a bright red colour. The action is reversible.

$$Hb + O_2 \rightleftharpoons HbO_2$$

Haemoglobin, purply red Oxyhaemoglobin, bright red

Combined with haemoglobin, oxygen is carried round the body in oxyhaemoglobin.

The haemoglobin combines 250 times more readily with carbon monoxide to form a much more stable compound carboxyhaemoglobin, which is very bright pink. In carbon monoxide poisoning (carbon monoxide forms 9·18 per cent of coal gas) the carbon monoxide replaces the oxygen, the carboxyhaemoglobin being carried round the body, and the tissues suffocate.

The envelope encasing this haemoglobin is a semi-permeable membrane. If corpuscles are placed in concentrated salt solution they shrink and become crenellated, plasmolysis taking place. If, on the other hand, the corpuscles are added to a weaker solution they take in water, swell up and burst. This is called the laking of corpuscles or haemolysis. It also occurs when certain foreign substances enter the blood.

In anaemia the number of red blood corpuscles is reduced and in one type of this disease (chlorosis) iron is given to help the body to build up more haemoglobin.

In the early embryo the red blood corpuscles are formed in the connective tissue; in the developing individual before birth they are formed in the liver and spleen, while shortly after birth new red blood corpuscles are formed in the bone marrow, though possibly some may still be formed in the spleen.

The worn-out red blood corpuscles are destroyed in the spleen and the broken-down red colouring matter goes to the liver, where it is excreted in the bile. No iron is excreted.

WHITE BLOOD CORPUSCLES
(LEUCOCYTES)

There are between 7,000 and 10,000 white blood corpuscles or leucocytes in 1 mm³ of normal blood. The number varies according to the
physiological state of the body, there being an increase, for example,
during digestion. The different types of leucocytes can be more readily
distinguished by staining, which shows up the nuclei and cell granules.
Some contain granules staining with an acid dye such as eosin, and are
therefore known as eosinophils; others contain granules which stain

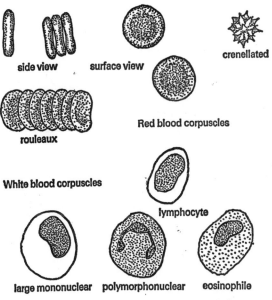

side view   surface view

crenellated

Red blood corpuscles

rouleaux

White blood corpuscles

lymphocyte

large mononuclear   polymorphonuclear   eosinophile

Fig. 46. Blood corpuscles (highly magnified).

with basic dyes, and are known as basophils; others, the neutrophils,
contain granules which stain with both basic and acidic dyes; others
contain no, or only very faint, granules.

Granulocytes. (1) Neutrophils or polymorphonuclear cells (about 65
per cent of the leucocytes). These have a nucleus divided up into lobes and
taking many forms, and the cytoplasm contains numerous small granules
staining with acid and basic dyes. These cells show amoeboid motion
and are capable of devouring bacteria, and probably also any injured
tissue. This phenomenon is known as phagocytosis. One of the body
reactions to inflammation is the increase in the number of these white
cells in the blood at the seat of injury or infection. Many of them die in
their fight against the invaders and form pus.

68   The blood

(2) Eosinophils (about 3 per cent of the leucocytes). These have a crescent-shaped or bilobed nucleus, and the cytoplasm contains coarse granules staining with acid dyes. These cells are not as a rule phagocytic and their function is unknown, although their number becomes increased in asthma and certain protozoon and worm infections.

(3) Basophils (about 0·5 per cent of the leucocytes). These have a bilobed nucleus and large coarse granules which stain with basic dyes such as methylene blue. These are rare in human blood, but occur in certain chronic infections.

Monocytes. (1) Mononuclear cells (1 to 2 per cent of the leucocytes). These are the largest cells in the blood. They have a large round or oval nucleus and vacuolated cytoplasm, which contains very faint granules. They engulf non-living particles, débris, and foreign material.

(2) Transitional cells (2 to 6 per cent of the leucocytes). These are slightly smaller than the mononuclears and have a lobed nucleus. They have, however, the same function as the larger monocytes and probably only represent a later stage.

Lymphocytes, large and small (about 25 per cent of the leucocytes). These have one large round nucleus and a small or large rim of cytoplasm according to the kind of lymphocyte. They are non-motile, but although they have no power to surround and ingest bacteria, they increase in number during certain special diseases such as tuberculosis and enteric fever. It is thought that they protect the body in some way by producing substances which destroy bacteria. They are temporarily increased in number after a fatty meal and exercise.

Formation of new leucocytes and destruction of the old. Lymphocytes are formed in the lymphatic glands and lymph nodules (see Chapter 9); neutrophils and eosinophils probably in bone marrow; and the large monocytes probably from the fixed histiocytes of the reticulo-endothelial system. It is thought that the worn-out leucocytes break up into fragments and may be removed by the active leucocytes.

✱ Blood platelets   (BLOOD CLOTTING)

These are small cells about one-third the size of the red blood corpuscles. There are about 300,000 per mm^3 of shed blood, and they are thought to be concerned with blood clotting.

Functions of the blood

Carriage of oxygen and the carriage of carbon dioxide. The blood carries oxygen round the body in combination with haemoglobin. While the

CARRIES FOOD REG/BODY TEMP.
NITROGENOUS WASTE
HORMONES
PROTECTION AGAINST DISEASE
HO₂ CO₂

BLOOD
DOES.

blood is in the capillaries of the lungs the oxygen combines with the haemoglobin inside the red blood corpuscles to form oxyhaemoglobin. When the blood reaches the capillaries of the organs, the oxygen is given up and haemoglobin is re-formed. The carbon dioxide passes back into the plasma. Here, by the addition of water, it is converted into carbonic acid. The carbonic acid then dissociates to provide bicarbonate ions. These bicarbonate ions (HCO_3^-) together with the sodium ions (Na^+), already present in the plasma, form sodium bicarbonate ($NaHCO_3$). It is as sodium bicarbonate that most of the carbon dioxide is carried from the tissues to the lungs where it is given up by a reversal of the changes described above. The mechanism of the interchange of oxygen and carbon dioxide will be described in the chapter on respiration.

Carriage of food. The blood carries food round the body to all the living tissues. This will be considered in more detail in Chapter 13.

Carriage of nitrogenous waste. During the 'wear and tear' of the body a certain amount of protoplasm is broken down. This forms nitrogenous waste, which passes in the blood to the liver to be converted into urea. This urea, along with the urea the liver forms from the breaking down of unused amino acids, passes in the blood to the kidneys, where it is excreted. The blood also carries certain other waste nitrogenous substances, e.g. creatinine, uric acid (see page 145), to the kidney.

Carriage of hormones or chemical messengers. A hormone is produced in one organ and passes in the blood to another organ whose activity it affects, e.g. when semi-digested food enters the duodenum a hormone called secretin passes into the blood. It goes in the blood to the pancreas and causes this organ to secrete its juice (see page 126).

Protection of the body against disease. It has already been stated that some of the white blood corpuscles show amoeboid motion and can engulf germs which invade the body. Also other white blood corpuscles and the blood plasma can produce antibodies, substances which destroy the germs or counteract poisons produced by them.

Regulation of body temperature. The manner in which the blood helps to regulate the temperature of the body will be discussed in Chapter 15.

Clotting of blood

If a blood vessel is punctured, although bleeding may go on for some time, eventually the blood becomes sticky and sets to a red jelly and forms a clot. The clot contracts and a pale yellow liquid or serum is squeezed out. Eventually the clot forms a scab, which protects the wound until it is healed.

When a clot is examined with a microscope it is found to consist of a network of threads with a number of red blood corpuscles in the meshes. These threads are composed of a protein, fibrin. It has been shown that when a blood vessel is injured the soluble fibrinogen in the blood changes into threads of insoluble fibrin. If blood is immediately whisked with a bunch of twigs when taken from the body, threads of fibrin form on the twigs and no clotting takes place. This kind of blood is called defibrinated blood.

The change from fibrinogen to fibrin is brought about by an enzyme (see page 116), thrombin.

$$\left(\begin{array}{c} \text{fibrinogen} \\ \text{(in blood)} \end{array} \xrightarrow[\left(\text{thrombin}\right)]{} \begin{array}{c} \text{fibrin} \\ \text{(in clot)} \end{array} \right)$$

Thrombin is not present in normal blood, but it is produced from prothrombin, normally present, by the action of thromboplastin. The latter is produced when tissue cells are injured.

Calcium salts are also necessary, no clotting taking place in blood if any substance is added which removes calcium from solution, e.g. sodium oxalate. Thus before clotting can take place there must be:

(1) Calcium salts in solution in the plasma; these are normally present.

(2) Injured tissue cells and blood platelets to produce the thromboplastin.

(3) Prothrombin, normally present in the blood to produce thrombin when acted upon by thromboplastin.

(4) Fibrinogen, normally present in blood to produce fibrin when acted upon by thrombin.

In a normal person clotting takes from 7 to 8 minutes. It is hastened:

(1) By the presence of any rough surface on the wound. Thus a piece of lint on the wound aids clotting.

(2) By raising the temperature slightly. This hastens the chemical changes involved in the formation of a clot.

(3) By applying snakes' venom.

(4) By decreasing the rate of blood flow through the vessel, e.g. by pressure.

Clotting is delayed:

(1) By keeping the blood in very smooth vessels (outside the body).

(2) By cooling.

(3) By addition of a soluble oxalate salt which removes the calcium in the form of insoluble calcium oxalate.

Clotting sometimes takes place under abnormal conditions in un-injured vessels (in the disease thrombosis). This is dangerous, as the clot formed may prevent the proper circulation of blood.

In the inherited disease haemophilia clotting is delayed. This delay is due to the congenital absence of a factor from the patient's plasma called antihaemophilic globulin (AHG). AHG is essential for normal blood coagulation. Patients may bleed extensively after slight cuts or other injuries – such bleeding may even be fatal. Where necessary clotting must be hastened by the intravenous injection of normal donor plasma containing AHG.

Blood groups

If blood from one animal is injected into the blood of a different animal, first agglutination and then laking of the red blood corpuscles takes place. It is the serum of one blood which acts on the blood corpuscles of the other blood. Thus if a drop of human blood is mixed on a slide with some rabbit serum the agglutination can easily be seen with the naked eye.

If the blood from one human being is injected into the blood of an-other human being agglutination followed by laking may or may not occur. This depends on the types of antigens the red blood corpuscles contain, and the types of antibodies the plasmas naturally contain. It has been found that the blood of man can be divided into four groups, A, B, AB, and O.

The ABO group system

	Red Cells	Plasma
Group A	Antigen A	Antibody b
Group B	Antigen B	Antibody a
Group AB	Antigen A Antigen B	Neither Antibody
Group O	Neither Antigen	Antibody a and Antibody b

Agglutination takes place when antigens and antibodies of the same type are brought together. Thus, for example, if Group A blood and Group B blood are mixed agglutination takes place. In blood transfusions it is necessary to be sure that the blood corpuscles of the donor will not be agglutinated by the serum of the recipient, for it has been proved by test-tube experiments that it is the corpuscles of the small volume which are agglutinated by the plasma of the large volume.

Recipient (*patient*)	Donor
Group A may receive blood from	Groups A and O
Group B	Groups B and O
Group AB	Groups A, B, AB and O
Group O	Group O

Patients of Group AB which can receive blood of all four groups are called universal recipients, while donors of Group O which can give blood to all four groups are called universal donors.

The percentages of the normal population belonging to the various groups are:

Group A	42 per cent
Group B	9
Group AB	3
Group O	46

The Rh factor

The four blood groups A, B, AB and O are not the only ones which should be taken into account when choosing a donor for transfusion. Up to 1940 some deaths which occurred after a blood transfusion remained unexplained. The explanation came during the course of animal experiments. Blood from the Rhesus monkey was injected into a rabbit and caused in the latter the formation of antibodies, for serum removed from the rabbit agglutinated the monkey's red blood cells. But this serum also agglutinated the red blood cells of 85 per cent of human beings showing that 85 per cent of human beings possessed the monkey's antigens on their red corpuscles. These persons were called Rhesus positive (Rh + for short); the remaining 15 per cent without the antigens were called Rhesus negative, or Rh—.

When the blood of a person who is Rh+ is injected into the circulation of a person who is Rh— the antibody is produced; unlike the antibodies in the ABO groups, it is not present naturally. As a result, the first transfusion causes no agglutination, but it is the second transfusion which may be fatal.

If an unborn child contains antigen Rh+ inherited from its father and the mother's blood is Rh—, some antigen may escape into the maternal circulation and stimulate the formation of an antibody. The antibody may then diffuse into the foetal circulation and destroy the red cells. This does not commonly happen and several stimuli are needed to provoke the formation of the antibody in the mother.

PRACTICAL

1 Prick the finger behind the nail, after tying a handkerchief round the base of the finger and bending the finger as much as possible. A drop of blood will exude. Make a blood smear on a slide and examine with miscroscope.

2 Add a drop of blood to several drops of 2 per cent salt solution on a slide. Examine with the microscope.

3 Add a drop of blood to several drops of pure water on a slide.

4 Add a drop of blood to a porcelain dish, cover with damp filter paper. After a few minutes remove paper and note jelly-like clot. Tease out on slide and examine with microscope.

5 Obtain two lots of blood from the butcher, one lot of which was whipped as soon as it left the animal to produce defibrinated blood. The second lot of blood clots.

   (a) Note the shreds of white fibrous material on the twigs. Examine under the microscope.

   (b) Add potassium oxalate to some of the defibrinated blood. Note the formation of a precipitate of calcium oxalate.

   (c) Note the serum rising above the clot in the second lot. Cut open clot and note appearance.

6 Examine a prepared stained blood film to show white blood corpuscles.

Additional experiments which can be performed

Details must be obtained from a practical physiology book.

1 Testing for blood proteins.

2 Testing for blood sugar.

3 Blood count.

SERUM AGGLUTINATED RED BLOOD CELLS
(ANTIGENS ON RED CORPOSCLES)
= (RHESUS POSITIVE)
(RH+)

B |Rh|D|Pos|

7 The heart

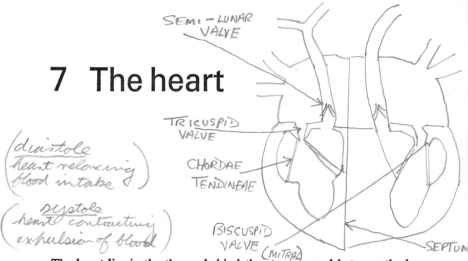

SEMI – LUNAR VALVE

TRICUSPID VALVE

CHORDAE TENDINEAE

BISCUSPID VALVE (MITRAL)

SEPTUM

(diastole
heart relaxing
blood intake)

(systole
heart contracting
expulsion of blood)

The heart lies in the thorax behind the sternum and between the lungs. It is a pear-shaped organ, the pointed end being towards the left of the middle line under the fifth left intercostal space (see Fig. 25). It is surrounded by a space called the pericardial cavity. The wall of this cavity, the pericardium, is attached to the heart, at the top around the large arteries and veins, and at the base to the diaphragm. It bends back over the surface of the heart to form its outer membrane, the visceral pericardium.

The inner wall of the pericardial cavity is moistened with fluid so that the heart beats smoothly, and the pericardium, being of a fibrous nature, prevents the heart becoming over-extended with blood.

The heart is a hollow organ, weighing 8–9 ounces (about 230 g), and its wall is made of three layers.

(1) *The pericardium,* already mentioned, forming an outer serous membranous covering.

(2) *The myocardium,* the middle layer, made up of cardiac muscle fibres.

(3) *The endocardium,* a serous membrane lining the heart cavities.

The heart is divided into four cavities, a right and left ventricle, and a right and left atrium. The atria are thin walled, the ventricles thick walled. The left ventricle, which pumps the blood round the body, has much thicker walls than the right ventricle, which only pumps the blood to the lungs. The atria and ventricles are separated by the atrio-ventricular valves, the tricuspid valve consisting of three flaps between the right atrium and right ventricle, and the mitral valve consisting of two flaps between the left atrium and left ventricle.

The flaps forming the valves are extensions of endocardium which hang down into the ventricles. When the latter are full of blood the valves come together and close up the two atrio-ventricular openings; and even when the ventricular pressure goes up the valves do not become everted back into the atria, for the edges of the valves are attached by slender thread-like tendons, the chordae tendineae, to the papillary muscles. The latter are muscular projections on the inside of the walls of the ventricles.

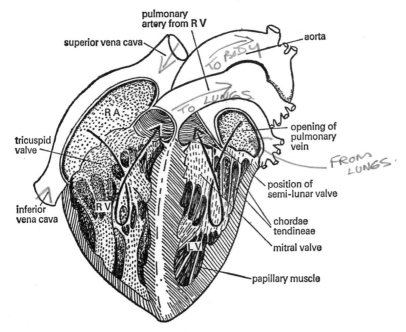

Fig. 47. Heart cut longitudinally.
RV, right ventricle. RA, right atrium. LV, left ventricle.

Two large veins, the inferior vena cava and the superior vena cava, enter the right atrium and bring blood back from all the body with the exception of the lungs. From the right atrium blood passes to the right ventricle and then out along the pulmonary artery to the lungs. The blood comes back from the lungs into the left side of the heart along the four pulmonary veins (two from each lung) which run into the left atrium. Blood passes to the left ventricle and then out via the aorta to all parts of the body except the lungs.

Between the right ventricle and the pulmonary artery and the left ventricle and the aorta are the semi-lunar valves. Each semi-lunar valve consists of three half-moon shaped flaps, with the more curved edge attached around the inside of the aorta and the straighter edge free, so that three pockets are formed. In the middle of each free edge is a fibro-cartilaginous nodule.

When the blood pressure in the artery is greater than that in the ventricles the valves close the orifice, their free edges come together in the form of a star, and the nodules are pressed together at the centre so that there is no reflux of blood into the ventricles.

In the right atrium near the entrance of the great veins is a collection of specialized cardiac muscle fibres forming the sinu-atrial node. The atria and ventricles are separated by fibrous non-conducting tissue in all but one place. At this place the gap is bridged by a bundle of highly conductive, modified heart muscle called the bundle of His. This starts at the atrio-ventricular node which has the same structure as the sinu-atrial node. The bundle passes in the septum between the two ventricles and then divides into two branches, each of which divides up in the ventricular muscle.

Heart beat

The contraction and relaxation of the heart is called the heart beat. For about 0·4 second the heart is relaxed with the atrio-ventricular valves open, but the semi-lunar valves closed. Blood rushes into the two atria and on into the two ventricles. The atria both contract and push more blood into the ventricles. Eddies are set up in the ventricles which cause the atrio-ventricular valves to float up and close the openings. The atria relax and the ventricles contract and the ventricular pressure goes up. The atrio-ventricular valves are kept from being pushed back into the atrium by the contraction of the papillary muscles.

Eventually the ventricular pressure becomes greater than the pressure in the aorta and pulmonary artery, the semi-lunar valves open and blood leaves the ventricles.

Blood pours out of the ventricles until the ventricles relax and the ventricular pressure falls below the aortic pressure and the pressure in the pulmonary artery. Then the semi-lunar valves close; a fraction of a second later the pressure becomes less in the ventricles than the atria, and the atrio-ventricular valves open and the whole heart relaxes.

A heart beat lasts about 0·8 second.

Contraction of atrium (atrial systole)   = 0·1 sec
„   „   ventricle (ventricular systole) = 0·3
Relaxation (diastole of the whole heart)   = 0·4

When the heart rate varies it is the length of the diastolic period which changes.

The heart beat is myogenic, that is, contraction and relaxation is a property of the heart muscle and does not depend on any connection with the central nervous system. This has been proved in many ways; for example, in a developing chick the heart begins to beat before nerves have grown into this organ. Also strips of frog's heart containing no nervous tissue will beat indefinitely in a solution of certain salts. The heart beat is, however, under the control of the nervous system, and regulated by it to the needs of the body.

The heart beat originates in the sinu-atrial node, the pace-maker of the heart, spreads over the atria so that the two atria beat simultaneously. The contraction wave is then conducted by the bundle of His to the ventricles, which likewise contract simultaneously.

Heart sounds

If the heart is listened to with a stethoscope two sounds can be heard at each beat: lu-ub, dŭp. The first lasts about one-third of a second, and corresponds to the contraction of the ventricles; the second is thought to be caused by the closure of the semi-lunar valves.

If the valves do not close the openings properly, as in certain heart diseases, and some blood leaks back, a gurgle, known in medicine as a murmur, is heard at the same time as the sounds.

Nervous control of the heart

This will be discussed in the next chapter.

PRACTICAL

1 Dissect a sheep's heart or ox's heart with pericardium attached and vessels left *in situ*.
2 Listen to the heart sounds with a stethoscope.
3 Observe the beating of a frog's heart in a pithed animal. The pithing should only be carried out by someone with experience.

8 Circulation

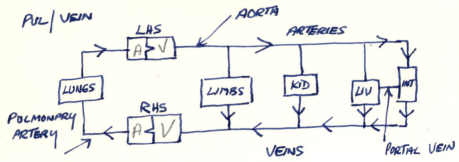

PUL/VEIN

AORTA

LHS

ARTERIES

A V

LUNGS

LIMBS

KID

LIV

INT

RHS

POLMONARY
ARTERY

A V

VEINS

PORTIAL VEIN

The path of the blood round the body

The general scheme of circulation was illustrated in Fig. 45 and described on page 64. The path of the arteries and veins will now be described in greater detail.

Systemic circulation

Arteries of the systemic circulation

The aorta leaves the left ventricle. Just beyond the semi-lunar valve it gives off the right and left coronary arteries, which supply the heart muscle. It ascends for a short distance and then curves backwards and to the left over the root of the left lung forming the aortic arch. The top of the arch is level with the third sterno-costal joint.

The aortic arch on the right gives off the innominate artery, which is only 2 in long and divides into the right subclavian artery and the right common carotid artery. A little farther on, on the left side, the arch gives off first the left common carotid and then the left subclavian artery.

The aorta then runs down the dorsal side of the body. In the thoracic region it is known as the thoracic aorta, lower down as the abdominal aorta. It gives off branches, the chief of which are:

(1) *The coeliac artery*, which arises just below the diaphragm. It divides into three branches, the hepatic artery which supplies the liver (see Fig. 77); the gastric artery to the stomach; and the splenic artery to the spleen.

(2) *The superior mesenteric artery* which arises about 1 cm below the coeliac axis.

(3) *The renal arteries*, which arise immediately below the superior mesenteric artery. These supply the kidneys.

(4) *The gonadal arteries* (spermatic in the male, ovarian in the female) to the reproductive organs. These arise below the renal arteries.

(5) *The inferior mesenteric artery*, which arises well below the renal arteries.

The superior mesenteric and inferior mesenteric arteries supply the mesentery and all parts of the intestines.

At the level of the fourth lumbar vertebra the abdominal aorta divides up into a right and a left common iliac artery. At the front of the sacrum each common iliac artery divides up into an internal iliac, which passes to the organs of the pelvis, and an external iliac, which enters the thigh as the femoral artery.

The femoral artery runs down the inside of the thigh and at the level of the knee it becomes the popliteal artery. The latter divides into two branches, the anterior tibial artery and the posterior tibial artery. The anterior tibial artery runs down the front of the tibia towards its inner side and into the foot where it becomes the dorsalis pedis. The latter gives off branches to supply the dorsal side of the foot and toes, and descends into the sole of the foot.

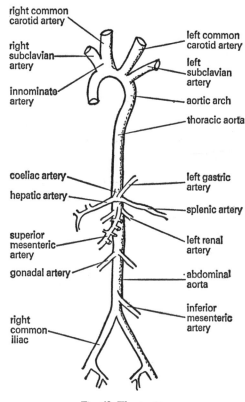

Fig. 48. The aorta.

The posterior tibial artery runs down behind the tibia embedded deeply in muscle. In the foot it divides into two branches, the lateral and medial plantar arteries. The lateral plantar artery receives the dorsalis pedis and gives off a branch which unites with a branch of the medial plantar artery to form the plantar arch. This arch gives off branches to the front part of the sole and the toes.

Branches of the aortic arch. The subclavian artery passes outwards, the right from the innominate artery, the left from the aorta. It passes over the first rib, under the clavicle, and enters the axilla as the axillary artery. Leaving the lower border of the axilla it runs down the upper arm as the brachial artery. At first it is towards the inner side, but gradually it passes to the front of the upper arm. About 1 cm below the elbow joint it divides into the radial and ulnar arteries.

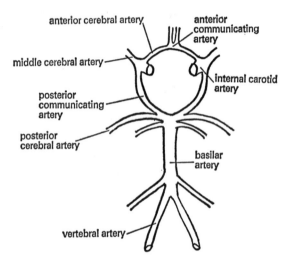

Fig. 49. The arteries at the base of the brain.

The radial artery passes along the radial side of the forearm, and the ulnar artery along the ulnar side to the wrist. They divide up into smaller branches to supply the hand. A superficial branch of the radial artery joins with the terminal part of the ulnar artery to form the superficial palmar arch. Similarly the terminal part of the radial artery joins with a deep branch of the ulnar artery to form a deep palmar arch. These two arches give off branches to the palm and the fingers.

The first part of the subclavian artery gives off the vertebral artery, which ascends the neck in the foramina of the transverse processes of the upper six cervical vertebrae, and then enters the skull to unite with the vessel of the opposite side to form the basilar artery. This divides into the right and left posterior cerebral arteries, which supply the hind part of the cerebral hemispheres. Each receives the posterior communicating branch from the internal carotid artery.

The right common carotid is a branch of the innominate and is confined to the neck. The left arises from the aortic arch and has a short thoracic part as well as the cervical part.

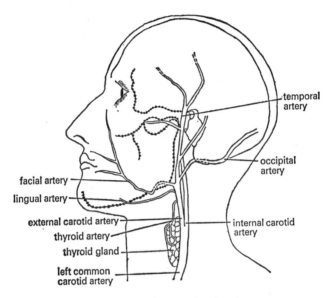

Fig. 50. Arteries of the head and neck.

In the neck at the level of the front part of the fourth cervical vertebra the common carotid divides into two branches, the external carotid and the internal carotid.

The internal carotid artery passes through the carotid canal in the temporal bone to enter the cranial cavity. It gives off the ophthalmic branch to supply the eye and then divides into two main branches, the anterior and middle cerebral arteries which supply the greater part of the cerebral hemispheres. Running backwards from the internal carotid artery is the posterior communicating artery which, as already mentioned, joins the posterior cerebral artery. At the beginning of the longitudinal fissure of the brain the two anterior cerebral arteries are joined by an anterior communicating artery about 4 mm in length. In this way a complete arterial circle known as the circle of Willis is formed within the cranial cavity.

The external carotid artery supplies the face and the outer side of the cranium. It has three main branches:

(1) *The occipital artery* which supplies the occipital region of the head.

(2) *The temporal artery* which supplies the skin and muscles outside the temporal bone and the side of the head.

(3) *The facial artery* which runs to the lower border of the mandible, runs over it to enter the face, where at the root of the nose it divides into a number of branches. It has a very tortuous course so that it is not stretched or damaged during the movement of the mandibles, lips, cheeks, etc.

In the above account only the main arteries are mentioned. They give off small branches so that every single organ in the body is supplied with blood. In the organs the arteries break up into smaller branches, the arterioles, and finally these break up into a network of capillaries which ramify through the organs. The capillaries join up to form small veins or venules, which unite to form larger and larger veins which take blood back to the heart. The course of the main veins will now be described.

Veins of the systemic circuit

Veins may be divided into the superficial, which lie just beneath the skin, and the deep. The latter or venae comitantes, which receive communicating branches from the superficial, run alongside the arteries and generally bear the same name. Sometimes an artery has two veins beside it, one on either side. In the lower limb the small veins from the toes and front part of the sole of the foot join up to form the venous plantar arch, and branches from this and the rest of the foot join up to form the two anterior tibial veins, and the two posterior tibial veins. They accompany the tibial arteries and join up at the lower border of the popliteal space to form the popliteal vein. The latter becomes the femoral vein, then the external iliac vein. This unites with the internal iliac vein to form the common iliac vein, which unites with its fellow of the opposite side to form the inferior vena cava.

The superficial veins of the legs are the long and short saphenous veins. The long

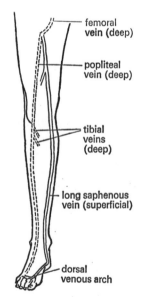

femoral vein (deep)

popliteal vein (deep)

tibial veins (deep)

long saphenous vein (superficial)

dorsal venous arch

Fig. 51. Deep and superficial veins of right leg (front view).

saphenous vein begins on the inner side of the foot by the joining together of small veins to form the dorsal venous arch. It passes up the inner side of the leg and enters the femoral vein.

The short saphenous vein begins on the outer side of the foot and passes behind the lower end of the tibia and runs up the back of the leg. It perforates the deep fascia in the popliteal space and enters the popliteal vein above the knee joint.

In the arm the superficial and deep palmar arches are formed from small veins from the palms and fingers, and branches from these and the rest of the palm join up to form two radial and two ulnar veins which accompany the arteries. In front of the elbow they unite to form two brachial veins, one on either side of the brachial artery. The brachial veins unite to form the axillary vein, which becomes the subclavian vein. The subclavian vein unites with the vein bringing back blood from the head to form the innominate vein.

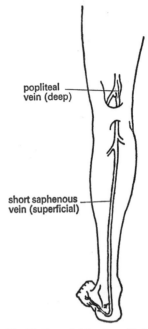

Fig. 52. Superficial veins of left leg (back view).

The most important superficial veins of the arm (fig. 53) formed from the network of veins in the hand are:

(1) *The cephalic vein* which runs on the lateral side of the forearm and upper arm. Below the elbow it gives off a branch to join the basilic vein. It eventually pierces the deep fascia in the shoulder to join the axillary vein.

(2) *The basilic vein* which runs on the medial side of the forearm and pierces the deep fascia in the upper arm to join the brachial vein.

(3) *The median vein* which ascends in front of the forearm and ends in the basilic vein. In a few individuals it divides below the elbow into two branches, one of which joins the basilic the other the cephalic.

In the dura mater, within the cranial cavity, are a number of venous sinuses, which are rather wide blood channels and drain blood from the cranial veins. These sinuses have tough membranous walls which do not easily collapse. The largest is the superior sagittal sinus in the fissure between the two halves of the cerebrum. Blood leaves the sinuses of the

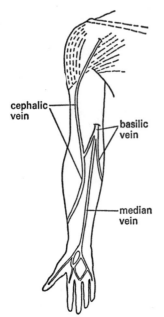

cephalic vein

basilic vein

median vein

Fig. 53. Superficial veins of right arm.

cranium along the emissory veins and enters the internal jugular vein. The vein also collects some blood from the superficial part of the face and neck. It runs down the neck and at the level of the clavicle unites with the subclavian to form the innominate vein. The greater part of the blood from the outside of the face, and also the deeper part of the face, is collected into the external jugular vein which runs down the neck and joins the subclavian vein.

The right innominate vein (about 1 inch in length) unites with the left innominate vein (about 2 in in length) to form the superior vena cava, which opens into the right atrium after receiving further branches from the neck and thorax.

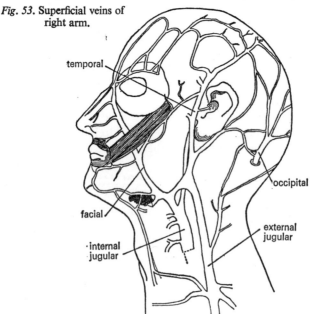

temporal

occipital

facial

external jugular

·internal jugular

Fig. 54. Veins of the neck and head.

The inferior vena cava formed by the union of the two iliac veins runs up the dorsal side of the body next to the aorta, receiving veins from the viscera and thorax. When passing on the dorsal side of the liver it receives the hepatic vein. It enters the right atrium.

Portal vein. The capillaries in the walls of the small intestine and the ascending and transverse colon join up to form the superior mesenteric vein. The capillaries in the spleen join up to form the splenic vein which receives the inferior mesenteric vein draining blood from the hind part of the large intestine.

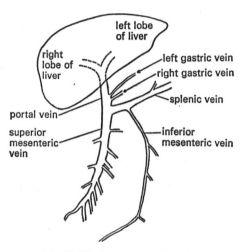

Fig. 55. The hepatic portal system.

The anterior mesenteric vein and splenic vein join to form the portal vein which receives two veins from the stomach and runs to the liver. There it breaks up into a second set of capillaries.

Pulmonary circulation

The pulmonary artery passes from the right ventricle upwards and backwards. At the level of the fifth thoracic vertebra it divides into a right pulmonary artery which goes to the right lung and left pulmonary artery which goes to the left lung. The pulmonary arteries when they reach the lungs break up into branches, which finally break up into capillaries lying against the walls of the air sacs in the lungs. The capillaries join up to form small veins, and eventually two pulmonary veins leave each lung. These pass to the left atrium.

Coronary circulation

The aorta gives off the right and left coronary arteries just after the semi-lunar valve. They divide and subdivide and finally break up into capillaries which ramify between the heart muscle fibres. The capillaries join up to form small veins, and these join up until finally a coronary sinus returns blood into the right atrium through an opening between the opening of the inferior vena cava and the right atrioventricular orifice.

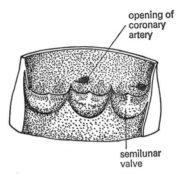

opening of
coronary
artery

semilunar
valve

Fig. 56. The aorta cut open.

Microscopic structure of the blood vessels

Arteries. These vessels have three coats:

(1) An outer fibrous coat, the *tunica adventitia*, whose function is protective.

(2) A middle muscular and elastic coat, the *tunica media*. In the aorta and large arteries there is more elastic than muscular tissue in this coat. In the small arteries and arterioles there is relatively more muscular tissue and far less, if any, elastic tissue. The blood supply to the organs is controlled by the contraction and relaxation of the muscular coat of the small arteries.

(3) An inner endothelial coat, the *tunica interna*. This is composed of flat pavement epithelium lying on a strong basement membrane, and, in the aorta and large arteries, a layer of elastic tissue. In health the smooth surface prevents clotting.

Capillaries. These have very thin walls consisting of a layer of endothelial cells one cell thick. Here and there surrounding the capillaries are branching cells called Rouget cells.

Veins. These resemble the arteries, but they have thinner walls and the lumen is larger. The walls have the same three coats, but the tunica media contains a great deal of white fibrous connective tissue in place of the muscular and elastic tissue. The veins are strong, but less muscular and elastic than the arteries. They do not have to withstand the strain of the heart beat.

The physiology of circulation

When the heart pumps blood into the aorta the blood already in the arteries cannot escape readily into the arterioles, since under normal conditions the arterioles are partially contracted. The pressure in the aorta and large arteries is therefore kept up (see table, page 88).

When the blood is pumped in at each beat the elastic wall of the first part of the aorta expands; during diastole this wall contracts to its original position and the blood is pushed on. The next part of the aorta then becomes swollen and a wave, known as the pulse, passes along the aorta and arteries at about 6–8 metres a second (about 18 times as fast as the blood inside the vessels). Where an artery runs on top of a hard structure and near the surface the pulse can be felt, a 'pressure point' being formed. Much can be learnt about the condition of the heart by examining this pulse.

In the small arteries which have non-elastic muscular walls there is a great fall in blood pressure (see table, page 88), the pressure being used up in overcoming friction. The walls not being capable of distension, the intermittent blood flow is here converted into a continuous flow, so that by the time the capillaries are reached the pulse has become obliterated.

In the veins which take the blood back to the heart the pressure is low, but the return of blood is helped by certain factors:

(1) The pressure of the skeletal muscles on the veins during any movement. Valves in the veins prevent the blood going in the opposite direction.

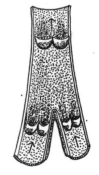

Fig. 57. Vein cut open to show valves (arrows indicate direction of blood flow).

When people stand a long time in one position they often begin to feel faint as the blood flows more slowly back to the heart. Thus muscular exercise is very important in promoting good circulation.

(2) The *vis a tergo* or unexpended force of the cardiac impulse from behind.

(3) The respiratory movements. During respiration the diaphragm descends and presses on the organs in the abdomen, and this presses the blood onwards. Also during the enlargement of the airtight thorax in respiration the pressure inside the thorax is reduced. Pressure is, therefore, taken off the large veins inside the thorax and blood is sucked into them.

Blood pressure

The pressure in the arteries varies during each heart beat. The highest blood pressure occurs during the heart systole and is called the systolic pressure; the pressure at the end of diastole is less and is called the diastolic pressure. The difference between the systolic pressure and the diastolic pressure is called the pulse pressure.

The arterial pressure may be measured by an instrument called a sphygmomanometer. It consists of a canvas armlet surrounding a broad rubber bag, which is connected to a pressure gauge and a syringe bulb. The armlet is strapped round the upper arm and air is pumped into the bag to raise the pressure. The canvas bag does not stretch, and the pressure of the distending rubber bag is exerted on the arm and compresses the blood vessel. When the pressure exceeds diastolic pressure the wave passing down the artery begins to oscillate the mercury in the manometer, and an approximate diastolic pressure can be read off the scale. The pressure which obliterates the pulse completely is the systolic pressure.

The blood pressure in the various vessels is compared in the following table:

Vessel	Range of pressure in mm mercury[1]	
Aorta	100 –200	
Large artery	90 –170	Note large
Small artery	40 – 70	drop
Capillary	20 – 30	
Small vein	10 – 20	
Large vein	0 – 10	
Vein inside thorax	−5 – 0	

[1] Pressure is given relative to atmospheric pressure.

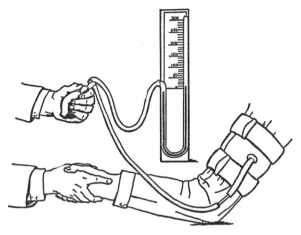

Fig. 58. Diagram of a sphygmomanometer.

Velocity of blood

The rate of the blood flow varies with:

(1) The amount of blood expelled from the heart in a given time. The speed and strength of the beat are regulated to the needs of the body by the nervous system.

(2) The total cross-sectional area of the vessels, the greater the cross-section the slower the flow. This can be understood if a river is watched flowing into a wider bed.

Now in the body:

The 'bed' of the large arteries ⟨ the 'bed' of the arterioles ⟨ 'bed' of the capillaries ⟩ the 'bed' of the small veins ⟩ the 'bed' of the large veins.

It follows that the blood flows slower and slower until it gets to the capillaries. The blood flow in the capillaries is slow and allows the interchange of materials between the blood and the tissues to take place. After leaving the capillaries the blood flow speeds up again.

There are sometimes some variations in the above scheme. For example, if arterioles of a certain organ dilate and there is a great decrease in pressure, more blood takes the path of the dilated vessels and therefore the velocity of the blood in that region goes up.

Control of the heart and circulation

The whole circulatory system is controlled to the needs of the body, not only by the nervous system but also by hormones in the blood.

D

(A general scheme of the nervous system should be known before this section is studied.)

Nervous control

The heart is supplied with two sets of nerve fibres. One set runs from the medulla oblongata in the vagus nerve, and these fibres form synapses with non-medullated fibres in the wall of the heart. Impulses normally pass down these fibres and act as a slight break on the heart rate. If in the frog they are stimulated the heart beats more slowly; if they are cut the heart begins to beat more rapidly.

The second set of fibres runs to the heart in the sympathetic system from the sympathetic ganglion at the foot of the neck. There these non-medullated fibres form synapses with sympathetic fibres which come from the central nervous system. If this second set is stimulated in the frog the heart beats more quickly.

It is thought the impulses affecting the rate of the heart act through the sinu-atrial node, the pace-maker of the heart (see page 76).

Heart reflexes

When the pressure in the aorta and carotid sinus, a small swelling at the beginning of the common carotid artery, becomes too high, the nerve endings in the walls of these vessels are stimulated and impulses pass up the vagus and glossopharyngeal nerves to the cardiac centre in the medulla oblongata. Impulses then pass down the vagus and slow the heart. At the same time impulses may pass to the arterioles causing them to relax. As a result the blood pressure is reduced and strained heart muscles, ruptured heart valves, and possibly a stroke from the tearing of the cerebral blood vessels, prevented.

When the pressure in the large veins goes up after the beginning of exercise and blood returns to the heart more quickly, nerve endings in their walls are stimulated; impulses pass up the vagus nerve to the cardiac centre in the medulla oblongata, and impulses come down the sympathetic system to the heart, and the impulses down the vagus are inhibited. The heart begins to beat more rapidly, and more blood which is needed during exercise passes through the body in a given time. Thus, if the pulse rate is counted immediately after exercise, it is above normal. It very soon goes back to normal as a result of the first heart reflex described.

Control of blood vessels

Non-medullated sympathetic nerve fibres pass to the arterial walls from the ganglia of the sympathetic chain in the thoracic region. There they form synapses with medullated fibres which come from the spinal cord. These in turn form synapses with fibres which pass down from the vaso-motor centre in the medulla oblongata.

Impulses which are normally passing down these fibres keep the arteries in a state of constriction and limit the blood flow through them. Thus if the cervical sympathetic nerve of a rabbit is cut the rabbit's ear flushes as a result of vaso-dilatation. When the upper cut end is stimulated the ear goes white as a result of vaso-constriction.

Some, probably all, organs are supplied with vaso-dilator fibres, stimulation of which causes the arteries to dilate. These fibres run in the cranial and spinal nerves.

The vaso-motor centre can be stimulated by the impulses passing in nerve fibres from the various organs, and by means of the vaso-dilator and vaso-constrictor nerves the blood supply of an organ can be regulated to its needs and the needs of the body as a whole.

After a meal the blood vessels of the alimentary canal relax and become full of blood, blood being directed away from the skeletal muscles by the greater contraction of their vessels. During muscular exercise, on the other hand, the skeletal muscles have a copious blood supply and the blood vessels of the alimentary canal are contracted. That is why exercise should not be taken after a large meal.

The vaso-motor nerves are also affected by emotion, e.g. blushing is caused by vaso-dilatation.

The capillaries as well as the small arteries may also be under the control of the nervous system. Probably impulses passing down nerve fibres act on the Rouget cells.

Chemical substances in the blood

Certain substances in the blood affect the heart and blood vessels. Adrenalin secreted by the suprarenal glands acts in the same way as the stimulation of the sympathetic system, speeding up the heart and constricting the arterioles. It is injected into the blood to counteract the symptoms of shock.

Excess carbon dioxide in the blood of the arterioles has the effect of increasing dilatation and so allowing more blood to flow through an active organ.

Histamine is a chemical substance which produces dilatation of the capillaries. It is formed when a tissue is injured. When the injury is widespread the capillary dilatation caused by the great amount of histamine passing into the blood may reduce the blood pressure so much, that a patient who has recovered from a primary shock caused by the inhibition of the nervous system may now suffer from secondary shock.

PRACTICAL

1 Examine the distribution of arteries and veins in a rabbit.

2 Examine prepared slides of T.S. of arteries, veins, capillaries.

3 Wrap the head of a tadpole in damp blotting-paper and place animal on slide. Observe circulation of blood through capillaries in tail.

4 Bind arm tightly. The part below bandage goes pale, bloodless and numb. Loosen bandage, the part below becomes flushed with blood. Bind bandage lightly, part below goes blue because the arteries are deep and remain open, and the superficial veins become constricted.

5 Count pulse:
   (a) sitting;
   (b) standing;
   (c) immediately after exercise; ·
   (d) 3 minutes after exercise.

6 Massage the arm downwards; swellings in the veins indicate the position of valves.

7 Measure blood pressure with sphygmomanometer.

9 The lymph and lymphatic system. The spleen

The blood flows through tissues in a network of small capillaries lined by a single layer of living cells which separate the blood from the tissues. In no organ, except the spleen, does the blood come in direct contact with the cells which it serves, and it is only in a few organs—for example, the liver—that all cells touch the outer surface of a capillary (or sinusoid).

In some tissues, e.g. cartilage, a living cell may be a considerable distance from a capillary. Every tissue is, however, permeated with a liquid called tissue fluid, and this liquid forms the 'middleman' between the blood and the cells. It carries food and oxygen to, and waste products away from, these cells.

Tissue fluid is the liquid which filters through the capillary walls into the surrounding tissue, normally the capillary walls allowing the passage of all the constituents of the plasma except protein. When the capillaries are fully open, motile leucocytes may pass out of the vessels between the individual cells of the capillary wall at junctional areas. The means by which other substances cross the vessel wall are many, some of which are not fully understood.

Tissue fluid is a slightly yellow, transparent and alkaline liquid containing less protein than the plasma. Otherwise it is similar in composition and it clots in the same way after it has left the vessels.

The amount of fluid filtering from the blood into the tissue spaces varies with the blood pressure which helps the process, and with the osmotic pressure of the non-filtering blood proteins which hinders the process. Normally the two pressures practically balance so that no movement of liquid takes place and the amount of liquid in the tissue spaces remains nearly constant. When this is the case the interchange of substances between the blood and the living cells takes place by simple diffusion through the fluid.

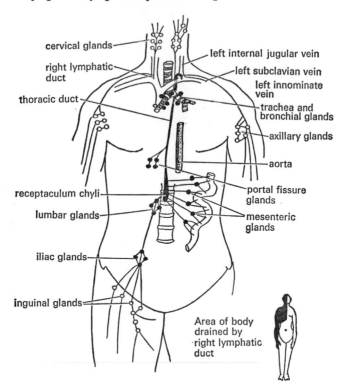

cervical glands

right lymphatic duct

thoracic duct

receptaculum chyli

lumbar glands

iliac glands

inguinal glands

left internal jugular vein

left subclavian vein

left innominate vein

trachea and bronchial glands

axillary glands

aorta

portal fissure glands

mesenteric glands

Area of body drained by right lymphatic duct

Fig. 59. Thoracic duct and position of main groups of lymphatic glands.

During tissue activity when a great number of chemical substances are produced, the osmotic pressure of the fluid in the tissue spaces rises considerably, and the capillary pressure goes up, with the result that there is great exudation of liquid. When this happens it has to be drained away.

From the tissue spaces the fluid which is now called lymph collects up into a second system of vessels, the lymph capillaries, which either end blindly as the lacteals in the villi of the small intestine (see page 125), or form a network of vessels throughout the tissue. These capillaries unite to form larger vessels, the lymphatics, which all run towards the chest. On their way they pass through the lymph glands, which will be described later in the chapter.

The lymphatics from the right side of the head and neck, the right arm, and the right side of the thorax, join to form a small duct about 1 cm long. This is called the right lymphatic duct, and it enters the blood system at the junction of the right subclavian and the right internal jugular vein.

The lymphatics from the legs join the receptaculum chyli, a sac-like structure at the beginning of the thoracic duct, the main lymph vessel of the body. The receptaculum chyli lies in front of the first and second lumbar vertebrae and drains lymph from the alimentary canal. The thoracic duct passes up through the abdomen and thorax to enter the blood system at the junction of the left internal jugular vein with the left subclavian vein. It drains lymph from all parts of the body, except the small area drained by the right duct. The lymph flows back to the thorax as the pressure in the tissue spaces is greater than the pressure in the thoracic duct and also as the muscles compress the vessels during movement. The lymphatic vessels contain valves which allow movement in one direction only.

If the absorption of lymph into the vessels from the tissue spaces is hindered in any way the lymph collects and the organ becomes swollen, the condition being known as oedema.

Certain substances injected into the blood cause an increased filtration of lymph into the tissues. These are known as lymphagogues, e.g. extracts from crayfish, mussel, etc. Nettlerash, which is due to an increased exudation of lymph into the dermis, is produced in some people after eating shell-fish. Histamine is produced by all injured tissues and is the cause of a blister being produced if the skin is burnt or rubbed by rough material; the liquid of a blister is lymph.

Lymph gland

A lymph gland is a kidney-shaped structure covered with a fibrous capsule. From the capsule, fibrous trabeculae run inwards, and these divide the organ up into a number of compartments. The compartments are subdivided into lymph spaces lined with endothelial cells.

An artery enters the gland at the hilum, it divides up into branches which run to the compartments and finally break up into capillaries. These join up to form veins, which join up until one vein leaves the organ alongside the artery.

Around the capillaries and filling the lymph spaces are masses of dividing lymphocyte cells. The spaces also contain some reticulo-endothelial cells.

Lymph flows into the gland along a number of afferent lymphatic vessels, it then percolates through the lymph spaces and collects newly

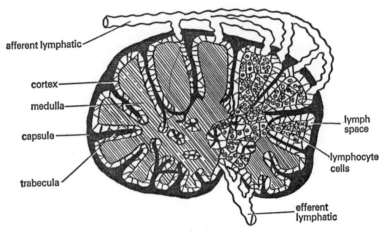

afferent lymphatic

cortex

medulla

capsule

trabecula

lymph space

lymphocyte cells

efferent lymphatic

Fig. 60. Section of a lymph gland.

formed lymphocytes from the outside of the masses. The lymph then passes out along the efferent lymphatic vessels which leave the gland at the hilum.

A lymph gland has two functions. First it produces new lymphocytes, and secondly the reticulo-endothelial cells remove the bacteria which the lymph conveys to the gland from the tissue spaces.

Smaller patches of lymphoid tissue are also found in the body, e.g. the adenoids, tonsils, Peyer's patches in the alimentary canal wall.

The spleen

The spleen is a dark purplish organ about 5 in long and 3 in wide. It lies in the left hypochondriac region of the abdomen, between the fundus of the stomach and the duodenum. It touches the diaphragm, the end of the pancreas, and the left kidney.

The spleen is covered on the outside by a capsule of muscular fibrous and elastic tissue. Into the inside of the organ, which is soft and spongy, the capsule sends trabeculae which branch to form a framework. Between the branches is the splenic pulp, consisting of a network of fibrous connective tissue with red blood corpuscles, lymphocytes and large reticulo-endothelial cells in the meshes. The reticulo-endothelial cells often contain broken-down red cells.

The lymphocytes are often collected together round branches of the splenic artery to form white masses called Malpighian corpuscles.

The splenic artery, a branch of the coeliac axis, enters the spleen and divides up into branches. The latter open directly into the pulp through which the blood percolates. The blood then collects up into small veins, which join to form the splenic vein. The splenic vein takes blood away from the organ and runs into the hepatic portal vein.

The flow of blood through the spleen is helped by the contraction and relaxation of the muscle fibres in the capsule and the trabeculae.

Functions of the spleen

(1) The worn-out red blood corpuscles are ingested and broken up by the reticulo-endothelial cells. The broken-down colouring matter goes to the liver, where it is converted into bile pigments.

(2) The spleen stores some of the iron from the haemoglobin of the broken-down red blood corpuscles.

(3) New lymphocytes are formed by division of cells in the Malpighian corpuscles.

(4) New red blood corpuscles are formed in the spleen during foetal life and sometimes during childhood after excessive haemorrhage.

(5) The newly formed lymphocytes can form antibodies to counteract poisons, so the spleen helps in the defence against disease.

(6) In some animals, e.g. cat, dog, but not in man, the spleen is an important reservoir of red blood cells and plasma.

PRACTICAL

1 Note the position of the spleen in a dissected rabbit.

2 Examine slides of the spleen and lymph glands under the microscope.

3 Examine slide of T.S. of small intestine to see lacteals in the villi.

4 Obtain an ox's spleen from the butcher. Cut open and examine. Wash it out under the tap to remove blood cells. Note the network of connective tissue.

10 Respiration

In order to function, living cells require energy and they obtain this energy by the oxidation of food; this process is spoken of as internal or tissue respiration. To enable this tissue respiration to take place oxygen must be taken into the body and the waste product, carbon dioxide, given out. This interchange of gas between the body and the outside atmosphere is spoken of as external respiration.

The function of respiration in the higher animals does not differ from that in Amoeba. In man the air passes in through the nostrils to the pharynx situated behind the mouth, then down the windpipe or trachea. The trachea divides up into two branches, the bronchi, each going to a lung. The air passes through these to the lungs, which contain a system of tubes and air sacs. When the air is in the lungs the interchange of gases between the atmosphere and the blood in the vessels in the lung walls takes place. The oxygen is carried round the body in combination with the haemoglobin of blood to the tissues which require it for their internal respiration. The carbon dioxide produced returns in the blood to the lungs, where it is exhaled.

Structure of the respiratory organs

On the inside of each nostril there are three bony projections known as conchae which are covered with a mucous membrane. This is highly vascular and has ciliated epithelial cells.

When the air passes up the nostrils it is warmed, moistened and purified by the small cilia which filter out dirt and germs. Thus in correct breathing through the nose pure, warm, moist air passes on to the lungs.

During infection, the mucous membrane of the nostrils becomes inflamed and the mucous secretion of the glands is increased.

In communication with the nasal cavities are the air cavities in the frontal, ethmoid, and sphenoid bone. These may also become inflamed in the common cold.

The nostrils open into the part of the pharynx (naso-pharynx) above the soft palate. Below is the oro-pharynx and the laryngeal pharynx, out

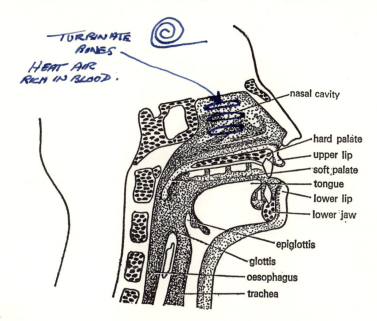

TURBINATE BONES

HEAT AIR RICH IN BLOOD.

nasal cavity

hard palate

upper lip

soft palate

tongue

lower lip

lower jaw

epiglottis

glottis

oesophagus

trachea

Fig. 61. Diagram to show upper part of respiratory tract.

of which open the oesophagus or gullet, and, in front, the trachea or windpipe. The opening into the trachea is called the glottis. During the process of swallowing the glottis is covered over by a flap, the epiglottis, which prevents food going down the 'wrong' way. It is impossible to swallow and breathe at the same time.

The first part of the trachea is swollen and forms the larynx or voice box. The whole trachea is about 4 in long and its wall is supported by incomplete rings of cartilage. These prevent the tube collapsing. The gap in the ring lies at the back of the trachea in contact with the oesophagus; this allows the wall to give slightly when food is swallowed.

The trachea and bronchi are covered inside by a layer of mucus. To this mucous layer will adhere foreign material such as bacteria and dust particles. The mucous sheet is constantly being moved upwards, ultimately to be expectorated, by the lashing movements made by the cilia of the respiratory epithelium.

At the level of the fifth thoracic vertebra the trachea divides into the two bronchi, which go to the lungs. The right bronchus is slightly shorter than the left. The bronchi are narrower than the trachea, but otherwise

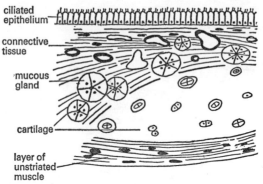

Fig. 62. Transverse section through part of the trachea.

similar in structure. When the bronchi reach the lungs they divide up again and again into small tubes called bronchioles, and each small bronchiole ends in an air sac.

The lungs

The lungs lie in the thorax, one on each side of the mediastinum. They are cone-shaped organs, the apex of the cone which is uppermost extend-

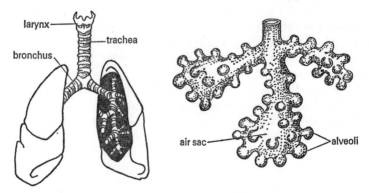

Fig. 63. The trachea, bronchi and lungs; and three air sacs (very highly magnified).

ing about one inch above the collar-bone. The base of each lung touches the diaphragm.

The right lung is divided into three lobes, the left into two (see Fig. 65). The bronchi, blood vessels, lymphatic vessels and nerves enter the lungs on their inner side forming their roots.

The inside of each half of the thorax is lined with a double serous membrane called the pleura. This membrane reaches on to the root

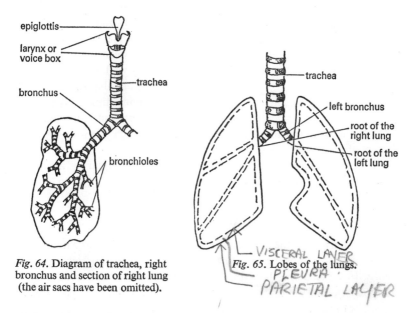

Fig. 64. Diagram of trachea, right bronchus and section of right lung (the air sacs have been omitted).

Fig. 65. Lobes of the lungs.

of the corresponding lung and covers the surface of the lung, from which it is inseparable. It passes into the fissures, dividing the lung into lobes.

The pleurae secrete a lubricating fluid enabling the lungs to move in the thorax without friction. In pleurisy the pleurae become inflamed, resulting in roughness of the surface, and movement of the lungs during respiration is very painful.

The large bronchiole tubes inside the lungs have the same structure as the bronchi, but the amount of cartilage continually decreases, so that eventually it is only found in small pieces.

As the bronchioles get smaller the muscular coat disappears, and the ciliated epithelium gives place to pavement epithelium. The elastic and white fibres still remain and bind the air sacs into lobules; the lobes therefore consist of many lobules. Each air sac has a honeycombed wall, the projecting pouches being called alveoli. An alveolus has a very thin wall consisting of one layer of pavement epithelial cells.

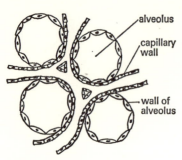

Fig. 66. Section of the lung wall (very highly magnified).

A pulmonary artery, containing deoxygenated blood, enters each lung, divides up into smaller branches which pass to the lobes and lobules. Finally a small branch breaks up into a network of capillaries which ramifies over the walls of each air sac and into the partitions between the alveoli. The thin capillary wall is fused with the alveolar wall, so that there is a very thin wall, consisting only of two layers of pavement epithelial cells, between the air in the air sacs and the blood. Through this wall the interchange of gases take place.

The capillaries join up to form branches of the pulmonary veins, which eventually form two pulmonary veins.

Mechanism of respiration

The two halves of the thorax are airtight cavities having a pressure slightly below atmospheric. During inspiration the thoracic cavity is enlarged by the dome-shaped diaphragm contracting and consequently going down, and the ribs and sternum coming up and out. The ribs are moved by the contraction of the intercostal muscles.

When the volume of the thoracic cavity increases, the pressure is reduced and is taken off the elastic walls of the lungs. The lungs enlarge, the pressure within them becomes reduced and air passes into them from the outside. This inspiration in helped by the widening of the glottis.

As a result of passive elastic recoil occurring in both the lungs and the chest wall, the thorax returns to its original size. This produces a relatively higher pressure in the lungs compared with that of the exterior, and air is therefore expelled. It is emphasized that normal quiet respiration is a passive process and that during it no actual contraction of the respiratory muscles occurs.

The breathing mechanism can be illustrated by a model described in the practical section.

In forced respiration the chest is enlarged to a greater extent; this is brought about by the assistance of other muscles, including those of the neck, shoulders, back and abdomen.

The respiratory movements are automatic, but they are regulated to the needs of the body by the nervous system.

Physiology of respiration

The amount of air which passes in and out of the lungs at each breath is known as the Tidal Volume, and varies in different persons and levels of activity between 250 and 500 ml. During a deep breath, an extra volume of air up to 2,000 ml and known as the Inspiratory Reserve Volume can be taken into the lungs. Also after normal expiration a further volume of air, about 1,000 ml, called the Expiratory Reserve Volume can be expelled.

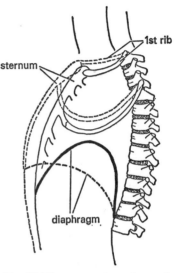

Fig. 67. Diagram to show shape of chest before inspiration (complete line) and after inspiration (dotted line).

The amount of air which can be breathed out after the deepest inspiration is known as a man's vital capacity (tidal volume + inspiratory reserve volume + expiratory reserve volume). In the average person it is about 4,000 ml, varying with the amount of exercise normally taken. In an athlete, for example, the vital capacity is much higher. It is necessary for air pilots to have a good vital capacity.

Even after forced expiration there is still about 1,250 ml of air left in the lungs. This is called the residual volume.

Inspired and expired air differ in composition, and the following table gives a rough comparison:

	Inspired air	*Expired air*
Oxygen	21% by volume	17% by volume
Nitrogen and inert gases	79%	79%
Carbon dioxide	A trace (0·03%)	4%
Water vapour	Varies in amount	Saturated with H_2O vapour
Temperature	Atmospheric temperature	Body temperature

It is seen that there is less oxygen in expired than inspired air and more carbon dioxide and water. The expired air contains a trace of organic matter.

The inspired air mixes by diffusion with the air in the lung tubes. The concentration of oxygen in the alveoli is greater than that in blood, and the oxygen diffuses into the red blood corpuscles, where it combines with the haemoglobin to form oxyhaemoglobin.

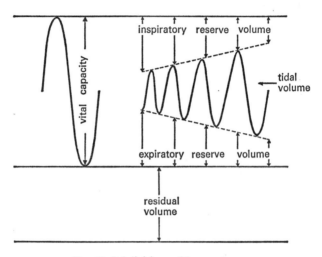

Fig. 68. Subdivisions of lung volumes.

Since the concentration of carbon dioxide in solution in the blood is greater than the concentration of carbon dioxide in the alveoli air, carbon dioxide diffuses out.

The nitrogen in the air sacs is in equilibrium with the small volume of nitrogen in solution in the plasma. No diffusion takes place, and therefore there is no difference in the nitrogen in inspired and expired air. The only function of the nitrogen is to dilute the oxygen.

Tissue respiration

The concentration of oxygen in the tissue spaces is very low. When blood reaches the tissue capillaries the oxygen passes out by diffusion and haemoglobin is reformed. The cells produce much carbon dioxide in internal respiration, and this diffuses back into the blood. The presence of the carbon dioxide hastens the breakdown of the oxyhaemoglobin.

The following table gives a comparison between the gases dissolved in 100 ml of venous blood and 100 ml of arterial blood:

	Venous blood	Arterial blood
Oxygen	8–12 ml	20 ml
CO_2	46 ml	40 ml
Nitrogen	1–2 ml	1–2 ml

These volumes were measured at 760 mm and 0°C.

In venous blood the loss of oxygen is greater than the increase of carbon dioxide, because in tissue respiration the oxygen unites not only with the carbon of the food to form carbon dioxide, but also with the hydrogen to form water, the amount used up in this way depending on the food substance oxidized.

The control of respiration

The rate of respiration varies with age. In a newly born baby it is about 40 times a minute; in a baby about 1 year old it is about 24 times a minute, while in adults it is 15–18 times a minute.

The movements are controlled by a respiratory centre, which lies in the medulla oblongata near the cardiac centre and the vaso-motor centre. If this centre is destroyed breathing stops, as the movements depend on the connection of the respiratory muscles to the central nervous system by nerves. The diaphragm is supplied by two phrenic nerves from the cervical plexus (see page 157), the intercostal muscles by nerves arising in the thoracic region of the spinal cord.

Respiratory movements are automatic. The lungs are supplied with branches of the vagus nerve containing sensory neurones. When the lungs collapse at expiration the nerve endings in the walls are stimulated and impulses pass up to the respiratory centre. These impulses pass down the nerves (phrenic and intercostals) to the respiratory muscles causing them to contract and inspiration is brought about. When the lungs expand at inspiration the same nerve endings are stimulated, impulses again pass to the respiratory centre, and inhibitory impulses pass down the nerves to the respiratory muscles causing them to relax and expiration is brought about.

This natural rhythm can only voluntarily be affected for short intervals of time; it is impossible to hold the breath for long.

The respiratory centre is influenced by other sensory neurones besides those supplying the lungs. In any irritation of the mucous lining of the nose violent expiration or sneezing takes place to remove the cause, while if something irritates the trachea coughing results.

A sudden application of pain to any part of the body produces inhibition of breathing for a short time. Cold water on the skin causes a deep breath to be taken.

The rate of breathing is adjusted to keep the percentage of carbon dioxide in the alveolar air at 5–6 per cent. During active exercise more carbon dioxide is produced and the percentage in the alveolar air tends to go up. The slight increase of carbon dioxide in the blood passing through the respiratory centre stimulates the centre and the rate of respiration is increased. In this way more oxygen required during active exercise is brought to the body. (It will be remembered that the rate of blood flow is increased during increased respiration [see page 88], and the percentage of carbon dioxide in the alveolar air is prevented from rising.)

Increased rate of breathing also takes place after the breath has been held. The respiratory centre is not only stimulated by increase of carbon dioxide in the blood but also by oxygen lack.

After forced breathing, when a great deal of carbon dioxide is removed through the lungs, the carbon dioxide concentration of the blood tends to fall, the respiratory centre is inhibited and breathing takes place more slowly until the concentration of carbon dioxide in the alveolar air again reaches 5–6 per cent.

When a human being breathes air containing a slightly higher percentage of carbon dioxide than normal, he does not feel any ill effect, but breathes more quickly to keep the percentage of carbon dioxide in the alveolar air constant. When the percentage rises above 10 per cent he begins to feel dizzy and loses consciousness.

Slight lack of oxygen in the inspired air has little effect, as the combining power of the haemoglobin is great. When the oxygen falls below about 13 per cent, however, great discomfort is felt and the skin goes blue. The respiratory centres are stimulated by lack of oxygen and excess of carbon dioxide, and convulsive respiratory movements take place. Death from oxygen lack or asphyxiation follows. Complete lack of oxygen, as in drowning, soon produces asphyxiation.

The effect of the lowered oxygen concentration is due to the reduced pressure of oxygen in the alveolar air. When climbing a mountain the atmospheric pressure decreases, and therefore the partial pressure of the oxygen goes down, although the percentage of oxygen in the air is the same as at sea level. Climbers suffer from this reduced pressure and frequently develop mountain sickness. Normal people, however, become acclimatized after a few days by increasing their respiratory movements and increasing the amount of haemoglobin in their blood.

Hypoxia (anoxia) is a general term indicating a lack of oxygen in the tissues. There are four possible types of hypoxia:

(1) *Hypoxic hypoxia.* The oxygenation of otherwise normal arterial blood is deficient. Examples of this type are inadequate ventilation of the lungs or insufficient oxygen in the inspired air.

(2) *Anaemic hypoxia.* Here the oxygen carrying capacity of the blood is reduced. The quantity of haemoglobin available for oxygen transport may be diminished in absolute amount, or as a result of combination with some substance such as carbon monoxide. In the latter instance the abnormal haemoglobin compound so formed is incapable of transporting oxygen.

(3) *Stagnant hypoxia.* The blood flow through the capillaries is reduced to such an extent that, in spite of normal oxygen content of the blood, the tissue requirements for oxygen cannot be met. This type may be localized as in the ears exposed to cold or generalized as in heart failure.

(4) *Histotoxic hypoxia.* The cells cannot use the oxygen delivered to them. This is the result of impairment of the enzyme mechanisms necessary for normal oxidative metabolism. Cyanides are potent poisons producing histotoxic hypoxia in this way.

Asphyxia is the term used to denote any condition where in addition to hypoxia there is also an increased level of carbon dioxide in the blood and tissues. Strangulation and suffocation are notable examples.

Cyanosis is the blue coloration of the skin and mucous membranes seen in the presence of inadequate oxygenation of the circulation blood.

The effects of hypoxia are many. Mild hypoxia may produce an euphoric state similar to alcohol intoxication whilst a more severe degree will result in coma. It must be emphasized that all episodes of hypoxia are damaging to the cells and tissues concerned. If exposure is anything more than transient the damage may well be permanent. The nervous system is especially vulnerable. At normal body temperature the brain is unable to tolerate total oxygen lack for more than three minutes without irreversible damage to the higher centres. Such damage may range from intellectual and psychological impairment to permanent coma. The patient with extensive cerebral damage may live as a 'vegetable' in coma for many months before eventually death occurs from a respiratory infection.

Artificial respiration

When the respiratory mechanism has failed as in carbon monoxide poisoning, air may be brought into the lungs artificially either by an operator alternatively increasing and reducing the capacity of the thorax, so causing air to be drawn into and expelled from the lungs, or as in the mouth to mouth method by an operator exhaling into the patient's mouth and so extending his lungs.

If artificial respiration is necessary for long periods the patient is laid, with his head protruding, into an airtight cabinet known as an iron lung operated by a motor, which rhythmically increases and decreases the size of the chest at the rate of normal respiration.

Voice and speech

Stretching across the cavity of the larynx from the front to back are folds of mucous membrane known as the vocal cords. The air currents from the lungs cause these to vibrate and the voice sound is produced.

The loudness of the sound varies with the force of the current, while the pitch of the sound varies with the length and tension of the cords. In children they are relatively short, producing a higher pitch than in adults.

During ordinary breathing the vocal cords are relaxed. Their tension and to some extent their length can be varied voluntarily by the muscles.

The sound produced by the vibration of the cords is modified in character by resonance as it passes through the nasal passages and the accessory air sinuses, thus giving tone to the voice. By means of the movements of the soft palate, tongue, lips and jaws the sounds of speech are produced.

PRACTICAL

1 Examine the lungs, windpipe, etc., of a sheep obtained from a butcher and in the body of a dissected rabbit.
2 Put glass tube in trachea and blow up lungs of the above specimen.
3 Examine slides of T.S. of trachea, nostrils, lungs, etc.
4 Model of chest. Set up the model as in Fig. 69. The rubber sheet representing the diaphragm can be pulled up and down at A, and the cavity representing the chest be made larger and smaller. Note the movement of the balloon (or lung).

5 Experiment to show that expired air contains more carbon dioxide than inspired air. Set up apparatus as in Fig. 70. Suck at A. This draws atmospheric air through the lime water in X. Blow at B. This sends air from the lungs through the lime water in X. Note the result. (When carbon dioxide is bubbled through lime water, the latter goes milky.)

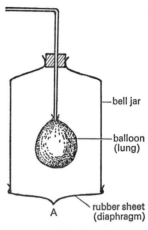

6 Breathe on to a mirror to show that expired air contains moisture.

7 Breathe into Jeyes' fluid. If the colour of the fluid becomes brownish the breath contains organic matter.

8 Measure the expansion of the chest during breathing. It should be about 3 inches. It can be increased by breathing exercises.

Fig. 69.

9 Measure tidal air and vital capacity in the following way. Fill a bell-jar with water and invert it in a sink full of water. Take a long right-angled glass tube and place one arm of the tube under the rim of the bell-jar. Take an ordinary breath and breathe out down the other arm of the tube. The breath will take the place of some of the water in the bell-jar. Move the position of the bell-jar until the level of the water is the same inside and out. Mark the level with sticky paper. Remove the bell-jar, invert it and add water up to the mark. The volume of the water which can be measured by adding it to a measuring cylinder gives the tidal air.

Repeat the experiment, taking as deep a breath as possible and breathing out forcibly. The volume measured gives the vital capacity.

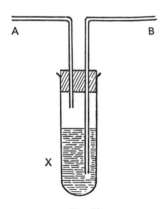
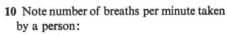

Fig. 70.

10 Note number of breaths per minute taken by a person:
   (*a*) sitting;
   (*b*) after exercise;
   (*c*) after forced breathing;
   (*d*) after holding breath.

11 Food

Every living cell requires energy and this energy is obtained from the oxidation or combustion of food. Food therefore has to be taken in to provide body fuel. Food also supplies the material for building up new protoplasm. This either replaces that broken down during the 'wear and tear' of the body or in young animals adds to the body, in other words the body grows.

At first sight the foods appear bewildering in number and variety, but they fall into three distinct classes:

(1) Carbohydrates.
(2) Fats and oils.
(3) Proteins.

Carbohydrates (e.g. starch, sugars, glycogen, cellulose)

These are compounds containing carbon, hydrogen and oxygen, the hydrogen and oxygen being in the same proportion as found in water. They are divided into three groups according to the size of the molecules:

(i) *Monosaccharides* or simple sugars, e.g. glucose or grape sugar and fructose or fruit sugar, both having the formula $C_6H_{12}O_6$. There are other less important, monosaccharide sugars containing fewer carbon atoms.

(ii) *Disaccharides* or complex sugars all having the formula $C_{12}H_{22}O_{11}$, e.g. sucrose or cane sugar, maltose or malt sugar, lactose or milk sugar.

(iii) *Polysaccharides*, e.g. starch, glycogen, cellulose. These have large molecules $(C_6H_{10}O_5)n$.

Starch, which is in the form of grains or granules, is manufactured in green leaves and is found as a store in many other parts of plants which are of economic importance, e.g. rice, maize, wheat (bread contains much starch), potatoes, etc. Starch is insoluble in cold water, and forms a paste in hot water. On boiling with dilute acid it is broken down by stages to glucose.

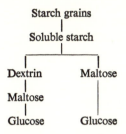

Test for starch: When iodine solution is added to starch or food containing starch a blue-black coloration is produced. Dextrin, the first stage in the breakdown of starch, gives a red-brown colour with iodine.

Glycogen, or animal starch, forms a white powder which is soluble in water. It is present in relatively large amounts in the liver and muscle tissues of animals.

Test for glycogen: When iodine solution is added to glycogen a port-wine colour is produced.

Cellulose forms the walls of most plant cells, and therefore is found in large amounts in fruit and vegetables.

When boiled with dilute acid it is broken down to glucose; the body, however, cannot digest cellulose, which forms what is called roughage in food.

Sugars. All the sugars are soluble in water.

Glucose or grape sugar occurs in grapes, many seeds, roots, honey, and also in blood.

Fructose or fruit sugar occurs in fruits, especially in tomatoes. It is found mixed with glucose in honey.

Sucrose or cane sugar is found in the sap of many plants, e.g. carrots, beetroots, sap of sugar-cane and sugar maple, bananas and other sweet fruits. Sucrose is a very valuable food, as it not only provides a good source of energy, but also stimulates appetite. On boiling with dilute acid it breaks down into glucose and fructose.

Lactose or milk sugar occurs in the milk of all mammals (see page 266). It is not very sweet, and if the lactose in milk were replaced by an equal quantity of sucrose, milk would be a very sickly food. On boiling with dilute acid it is converted into glucose and galactose.

Maltose or malt sugar is one of the breakdown products of starch. For this reason it is found in germinating grain. By boiling with acid it is converted into glucose.

All the sugars mentioned except sucrose contain a reducing group in their molecules and are known as reducing sugars. When they are boiled with alkaline solution of cupric oxide (Fehling's solution) they reduce it to the insoluble cuprous oxide, which forms a brick-red precipitate.

Sucrose contains no such group and is known as a non-reducing sugar. It will only act on Fehling's solution after it has been boiled with dilute acid to convert it into glucose and fructose.

The reducing property of some sugar is used as a basis for sugar tests (see Practical at end of chapter).

Proteins

All proteins are compounds containing carbon, hydrogen, oxygen, nitrogen and sulphur. They are essential constituents of animal food, as they are required for the building up of protoplasm and thus for growth and tissue repair. They may be used as a source of energy, although carbohydrates and fats are more suitable for this purpose.

The simple proteins are classified below according to their solubility.

(*a*) Albumins, e.g. serum albumin, egg albumin, some cereal proteins, lactalbumin from milk. These are soluble in water and dilute solutions of salts, acids, alkalis. They are coagulated by heat.

(*b*) Globulins, e.g. serum globulin, myosin of lean meat, legumin in beans, greater part of seed protein reserves. These are insoluble in water, but soluble in sodium chloride.

(*c*) Glutelins, e.g. glutelin in wheat. Glutelins are only found in plants. They are soluble in acids and alkalis, but insoluble in all neutral substances.

(*d*) Alcohol-soluble proteins, e.g. gliadin of wheat, zein from maize. Like the glutelins they are only found in plants.

(*e*) Skeletal proteins, e.g. keratin of horn, elastin from tendinous material. They are only dissolved by concentrated solutions of acids and alkalis.

Besides these simple proteins there are compound proteins in which the protein molecule is combined with some other substance, e.g. haemoglobin built up from a protein, globin, and a substance containing iron called haem; nucleoproteins, compounds built up from nucleic acid and one or more proteins, which contain phosphorus in addition to the other usual elements.

All proteins have large molecules, and if they are soluble form colloidal solutions. When boiled with acid they are broken down first into simpler nitrogenous compounds known as peptones, and eventually the soluble and diffusible amino acids with relatively simple molecules are produced. The very numerous proteins are built up of only 24 amino acids, not more than 20 of which are usually present, the remainder occurring only in a few special cases. Just as 26 letters of the alphabet can be joined together to form all the words of our language, so can 20 or so amino acids be united to form all the existing proteins.

Test for proteins. *Biuret test.* When excess caustic potash solution and two or three drops of copper sulphate solution are added to a 'solution' of a protein, a violet coloration is produced. This test may also be used for peptones. With these substances a pink coloration is produced.

Xanthoproteic test. When concentrated nitric acid is added to protein and the mixture heated, a yellow precipitate or solution is produced. On cooling and the addition of concentrated ammonia solution the colour changes to orange.

Millon's test. When Millon's solution is added to a protein and the mixture heated, a reddish precipitate or a slight reddish coloration is produced.

Fats and oils

These are compounds containing carbon, hydrogen and oxygen, but relatively less oxygen than a carbohydrate. They are neutral substances and break down on boiling with acid into glycerol and a fatty acid.

The most important fats in the animal body are palmitin, stearin and olein. The three are found in butter, and palmitin and stearin are both found in beef and mutton fat.

Palmitin and stearin melt at a higher temperature than olein, which is liquid at normal temperatures. The hardness of a fat depends on the proportion of each present.

Fats and oils are insoluble in water, but soluble in alcohol and ether. They may form an emulsion with water, an emulsion being a suspension of very small insoluble globules of one liquid in another, e.g. in cod-liver oil emulsion very small globules of cod-liver oil are suspended in water.

In animals fat is stored around the abdominal organs and in the subcutaneous tissue. In whale this subcutaneous fat forms a thick layer called blubber.

Fat is not only used in animals as a food reserve, but also helps to keep the animal warm.

Some fat stores in plants are of great economic importance, e.g. olive oil from the fleshy fruit of olive; castor oil from the castor-oil seed.

Test for fat. When an oil or fat is rubbed on a filter paper a translucent stain is produced. This oily stain is insoluble in water, but is soluble in ether and benzene. When a liquid known as Sudan III is added to the stain a salmon-pink coloration is produced.

Other essentials of a complete diet will be discussed in the Hygiene section of this book.

PRACTICAL

1 Test for starch. Add some iodine solution to powdered starch. Note coloration.

2 Test for sugars:
   (a) a reducing sugar. Dissolve a small quantity of glucose in water. Add an equal volume of Fehling's solution A and B and heat. A brick-red precipitate is produced.
   (b) a non-reducing sugar. Dissolve a small quantity of sucrose in water. Divide solution into two lots. To the first lot add an equal volume of Fehling's solution and heat. There is no precipitate produced. To the second lot add two or three drops of concentrated hydrochloric acid and heat. Then add an equal volume of Fehling's solution and heat. A brick-red precipitate is produced.

3 Test for proteins. Dilute white of egg with water, stir well, and use about 5 ml of the mixture for each of the following tests:
   Biuret test. Add an excess (about 10 ml) of caustic potash solution and 3 or 4 drops of copper sulphate solution. Note coloration.
   Xanthoproteic test. Add about 5 ml of concentrated nitric acid and heat. Note coloration. Cool test-tube under tap and add 5 ml or more of concentrated ammonia solution. Note coloration.
   Millon's test. Add about 5 ml of Millon's reagent and heat. Note coloration.

4 Test for fat. Place a drop of olive oil near a water stain on a filter paper. Note that the oily stain is more translucent than the watery stain. Show that the oily stain is soluble in ether and benzene, but insoluble in water. To

another oily stain and a water stain add Sudan III. Note that the oily stain is pinker than the water stain.

5 Test certain foodstuffs (e.g. potato, carrot, turnip, bread, flour, meat, cheese, fish, etc.) for starch, sugars, fats and proteins. The foodstuffs should be well chopped up or ground up in a mortar before testing. Record the results carefully.

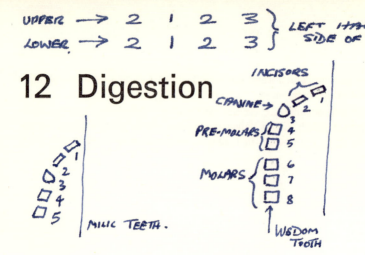

UPPER → 2 1 2 3 } LEFT HAND
LOWER → 2 1 2 3 } SIDE OF MOUTH

INCISORS
CANINE → 1, 2
PRE-MOLARS 3, 4, 5
MOLARS 6, 7, 8
↑ WISDOM TOOTH

MILK TEETH. 1, 2, 3, 4, 5

12 Digestion

Before food can be utilized by the body it has to be converted into soluble diffusible substances which can pass through the walls of the small intestine and blood vessels into the blood, and be conveyed round the body in the blood. This preparation of food for absorption is termed digestion, and takes place in the alimentary canal, a long muscular canal beginning at the mouth and ending at the anus.

During digestion all proteins are converted into amino acids, fats into fatty acids and glycerol, and all carbohydrates into the very simple carbohydrates, glucose, fructose and galactose. All these changes are brought about by substances called enzymes, contained in the secretions from the digestive glands.

Enzymes or ferments are found in both animals and plants. They bring about chemical changes without themselves becoming changed in the process, and are therefore known as biological catalysts. They are produced by the living protoplasm, though are not themselves alive, and they can work outside the living cells. For example, ptyalin, found in the saliva produced by the cells of the salivary glands, carries out its work in the mouth.

An enzyme is specific in its action; ptyalin, for example, will act on starch, but it has no effect on protein.

Enzymes are destroyed by heat and generally have a particular medium for optimal activity, e.g. ptyalin acts best in a slightly acid medium, its activity is destroyed by strong acid or alkaline solution.

The alimentary canal

Mouth

The mouth is an oval cavity with its opening to the outside, being surrounded by a pair of lips. It consists of two parts, the outer part or vestibule, which is the space outside the teeth and maxillary bone and within the lips and cheeks; and the inner part or true cavity of the mouth. This communicates with the oro-pharynx (see page 99). The roof of the

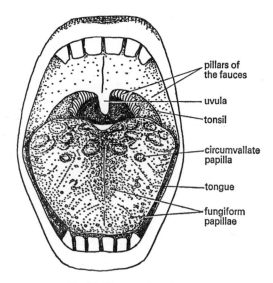

pillars of the fauces

uvula

tonsil

circumvallate papilla

tongue

fungiform papillae

Fig. 71. The mouth and tongue.

mouth consists of the palate; the front part is hard, being supported by the inner wings of the maxillary bones and the palate bones; the posterior part or soft palate consists of muscles and mucous membrane, and from its centre the uvula hangs down. Arching outwards and downwards from each side of the uvula are two curved folds of muscle and mucous membrane called the pillars of the fauces. The tonsils, patches of lymphoid tissue, lie between these.

The mucous membrane of the soft palate is continuous with the mucous membrane lining the mouth. This membrane has stratified epithelium on its outside.

The tongue lies on the floor of the mouth and is attached at the back to the hyoid bone. It is a muscular organ covered with papillae. These will be described in more detail on page 182.

Teeth

Each tooth consists of a crown, which is exposed, and one or more roots, which are embedded in sockets in the jawbones. Between the crown and the roots is a narrow neck.

The crown is covered with enamel, a very hard substance containing a great deal of phosphate and calcium carbonate. Within the enamel is a thick layer of bony substance called dentine, which extends up into the

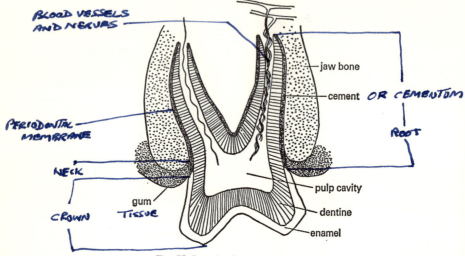

Fig. 72. Longitudinal section of tooth.

roots. Within the dentine is the pulp cavity full of connective tissue with nerves and blood vessels. There are little branching tubes in the dentine, through which nutritive material is conveyed from the blood in the vessels of the pulp cavity.

The roots are fixed to the fibrous tissue layer lining the socket by means of cement, a type of very hard bone.

Decay is brought about by bacteria, which multiply on small food particles, producing acid which destroys the enamel. The dentine is then attacked by the bacteria. When the nerves in the pulp cavity are irritated toothache develops.

There are two sets of teeth, the temporary and the permanent. In the temporary or milk set there are twenty teeth. On each jaw there are:

(1) 4 incisors or cutting teeth, 2 on each side of the mid-line. These are 'cut' at about 6 months, the central ones being cut first.

(2) 2 canines or eye teeth, 1 on each side of the mid-line. These are 'cut' at about 18 months.

(3) 4 molars, 2 on each side. These are the double teeth with more than one root. The first molars on each side appear at about 12 months. After the canines the last molars are cut.

The teeth in the lower jaw generally appear before the corresponding teeth in the upper jaw.

Between the ages of 5 and 10 the temporary teeth gradually fall out and are replaced by the permanent set. The teeth replacing the milk molars are called the pre-molars, as 12 other teeth, 3 on each side of both the upper and lower jaws, grow in behind. These are the molars of the permanent set. The first molars generally grow in at about 6 years, before the temporary teeth have fallen out. The second molars come at about 12 years and the last molars or wisdom teeth grow in at 18 plus.

Salivary glands

These are compound racemose glands (see page 13) divided up into lobules. Ducts from the lobules join up to form the main duct, which

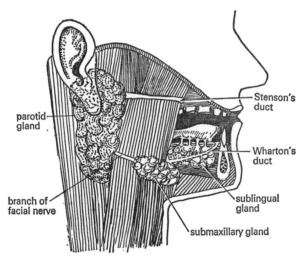

Fig. 73. The salivary glands.

leaves the gland and carries the saliva to the mouth. There are three pairs of salivary glands:

(1) *Parotid glands.* These are the largest glands and lie below and slightly in front of the ear. Each is connected with the mouth by Stenson's duct, which opens opposite the second molar on the inside of the cheek.

(2) *Submaxillary glands.* These are the next largest, and they lie beneath the mandible on each side, and the secretion from each gland is taken to the floor of the mouth along Wharton's duct.

(3) *Sublingual glands.* These lie beneath the tongue. They are the smallest and they pour their secretion through a number of small ducts into the floor of the mouth.

Digestion in the mouth

The food is divided up into small pieces or masticated by the teeth and well mixed with saliva, which flows out of the salivary glands. There are two secretions of saliva:

(1) A psychic flow due to a mental stimulus, e.g. the watering of the mouth at the sight, smell, or thought of food.

(2) A second flow when food is actually in the mouth.

Saliva is a colourless, slimy liquid, slightly alkaline in reaction. It is chiefly water containing in solution some salts, mucin, and the enzyme ptyalin. It has important digestive functions:

(1) It moistens and lubricates the food, so facilitating swallowing.

(2) It dissolves part of the food, making taste possible. This stimulates the secretion of more saliva and of other digestive fluids.

(3) The enzyme ptyalin begins the digestion of starch, converting cooked starch to dextrins and maltose or malt sugar.

The proper mastication of food is very important, as it exposes a greater surface to the saliva and the other juices secreted in later stages of digestion.

Pharynx and oesophagus

The mouth leads into a cavity known as the pharynx. It is a cone-shaped passage about 5 in long and divided in three parts:

(1) The naso-pharynx lying behind the nose. Into this part the Eustachian tube (see page 179), from the middle ear and the two nostrils, opens.

(2) The oro-pharynx, into which the mouth opens. The tonsils lie on the lateral walls of this part.

(3) The laryngeal pharynx, which extends behind the larynx to the level of the sixth cervical vertebra. Out of this part open the larynx and oesophagus, a tube leading from the pharynx to the stomach.

The pharynx and the oesophagus are both lined with a mucous membrane similar to that lining the mouth, with a thick layer of stratified epithelium. Beneath the epithelium there are numerous mucous glands which secrete mucus to keep the membrane moist. In a relaxed throat these glands are generally enlarged.

In the pharynx outside the mucous coat there is a fibrous connective tissue coat and beyond a muscular coat consisting chiefly of constrictor fibres. In the oesophagus the middle fibrous coat is lost but there are four coats:

(1) The inner mucous coat, already described.

(2) The submucous coat containing large blood vessels, nerves and more mucous glands.

(3) The muscular coat, consisting of a layer of longitudinal muscle fibres and a layer of circular muscle fibres.

(4) An outer fibrous connective tissue coat.

The oesophagus is about 9–10 in long, and lies between the trachea and the vertebral column. It passes through the thorax and goes through the diaphragm to enter the abdomen. It communicates with the stomach at the cardiac orifice.

Swallowing

The food is formed into a ball or bolus by the tongue and cheeks, and the tongue then passes it back between the pillars of the fauces. The pharynx then passes the food on to the oesophagus. At the same time the soft palate rises to cut off the posterior nares, the internal openings of the nasal cavities, and the glottis is drawn up and closed by the epiglottis, so that breathing temporarily ceases. After the food has left the back of the mouth it is no longer under the control of the will.

The food passes down the oesophagus by muscular action known as peristalsis. Peristalsis is a wave of relaxation followed by a wave of contraction, the circular muscle fibres being inhibited in front of the food and stimulated behind it.

Stomach

The stomach is a large bag situated in the abdomen immediately below the diaphragm. It is partly in the epigastric region of the abdomen, partly in the left hypochondriac region. The size varies with the amount of food it contains.

The stomach consists of two parts, a large vertical portion or body lying to the left, and a smaller transverse or pyloric portion below the body on the right.

The upper region of the body, which extends above the opening of the oesophagus, is called the fundus; the part around the cardiac opening is known as the cardiac region of the stomach.

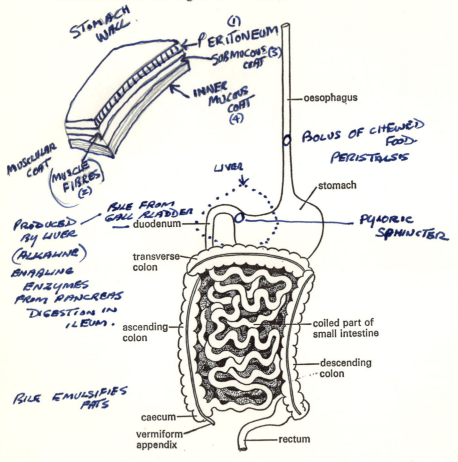

Fig. 74. Diagram of the alimentary canal.

The stomach wall consists of four coats:

(1) An outer serous coat (the peritoneum).

(2) A muscular coat consisting of, from without, longitudinal, circular and oblique unstriped muscle fibres.

(3) A submucous coat containing blood vessels in loose areolar tissue, which connects the muscular layer with –

(4) The inner mucous coat. This is thick and its surface is soft and

velvety. When the stomach is contracted this coat lies in folds, which disappear when the organ is distended.

When examined under a lens the inner surface of the mucous membrane is honeycombed in appearance, as it is covered with shallow depressions. These are the ducts of test-tube shaped gastric glands in the mucous membrane, and one or more of these glands discharge into each duct.

The epithelial cells covering the inside of the mucous membrane are columnar in shape. In the glands they are of two kinds, the chief or central cells and the larger oxyntic cells which are oval in shape. The glands secrete gastric juice, which is a clear, colourless, strongly acid fluid. Its acid reaction is due to the hydrochloric acid it contains. This acid is formed in the oxyntic cells from the sodium chloride they extract from the blood. The juice also contains certain solids in solution and two enzymes known as pepsin and rennin. Both are found in the chief cells.

Digestion in the stomach

The stomach receives the food from the oesophagus, and the presence of food in the stomach stimulates peristalsis, so that the food becomes well mixed and is passed on to the pyloric portion. The presence of food also stimulates the gastric glands to secrete their juice. This is the second flow of gastric juice, the first flow is a psychic flow taking place when food is in the mouth.

When the food first reaches the stomach salivary digestion is continued. When the food is mixed with the gastric juice the hydrochloric acid inhibits the action of the ptyalin and gastric digestion proper begins.

The functions of the gastric juice:

(1) The pepsin begins the digestion of proteins, converting some to peptones.

(2) The acid provides the necessary acid medium for the activity of the pepsin, and also kills any germs taken in in the food.

(3) The rennin clots milk, converting the soluble caseinogen in milk into insoluble casein. The clot can then be acted upon by the pepsin.

When the food has been converted into a semi-liquid mass called chyme, the pyloric sphincter separating the stomach from the duodenum, the first part of the small intestine, opens and some of the chyme passes on. The sphincter closes until the acid chyme passed on has been made alkaline by the fluids in the duodenum. It then reopens and allows more chyme to pass on. During a meal food is continually passing into the stomach, but only small quantities of chyme pass into the duodenum at any one time.

In the stomach the fat is melted and may be acted on by a fat splitting enzyme (lipase) in any liquid regurgitated from the duodenum. A gastric lipase may be contained in the gastric juice.

The stomach, like other organs, requires periods of rest, so that meals should be eaten at regular intervals, and small snacks of food should not be eaten continuously throughout the day.

Anti-anaemic factor. The stomach secretes a substance known as the gastric or intrinsic factor which acts on a certain substance, known as the extrinsic factor, in certain foods such as beef and other meats to form the anti-anaemic principle. This passes in the blood to the liver where it is stored (see page 136).

Vomiting

When any food in the stomach irritates the walls, or when other abdominal organs are irritated, the contents of the stomach are expelled out of the mouth. This takes place by the compression of the stomach by a violent respiratory movement. During the process the glottis is shut by the epiglottis and the pyloric sphincter is closed.

Small intestine

The small intestine is about 20 ft long and consists of two parts, the duodenum and the coiled part.

The duodenum is a short curved tube about 10 in long, circling round the pancreas. It has no mesentery. Opening into it about 4 in from the pyloric sphincter is a duct which is formed by the union of the bile duct from the liver and gall bladder, and the pancreatic duct from the pancreas.

The long coiled part of the small intestine lies in the central and lower part of the abdominal cavity, and is attached by mesentery to the posterior wall of the abdomen. In the mesentery, branches of the mesenteric artery and vein are supported. The upper two-fifths of the coiled part is called the jejunum, the last three-fifths the ileum.

The wall of the small intestine consists of the same four coats as the stomach:

(1) The outer serous coat (the peritoneum).

(2) The muscular coat, with longitudinal fibres on the outside and circular fibres within. There are no oblique fibres.

(3) The submucous coat.

(4) The inner mucous coat, which in the small intestine is permanently folded. The folds are called valvulae conniventes, and they increase the

area of secretion and absorption, two processes which go on in the small intestine. They also prevent the food passing down too quickly.

The folds are covered with minute hair-like projections called villi, which give the surface a velvety appearance. A villus contains a number

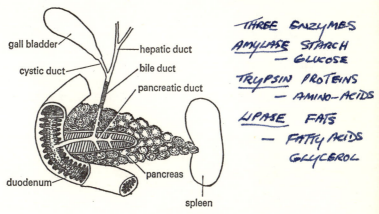

gall bladder
hepatic duct
cystic duct
bile duct
pancreatic duct
duodenum
pancreas
spleen

THREE ENZYMES
AMYLASE STARCH
— GLUKOSE
TRYPSIN PROTEINS
— AMINO-ACIDS
LIPASE FATS
— FATTY ACIDS
GLYCEROL

Fig. 75. Duodenum, pancreas, gall bladder and spleen.

of blood capillaries, muscle fibres and a central lacteal supported by connective tissue.

There are different kinds of glands in the wall of the small intestine:

(1) *Glands of Lieberkuhn.* These are test-tube shaped glands, which open directly to the surface and secrete the intestinal juice. Their orifices can easily be seen with the aid of a hand lens between the villi. They are formed by the mucous membrane.

(2) *Brunner's glands.* These are only found in the duodenum, and are situated in the submucous coat where they branch. They open by a duct on the inner surface of the duodenum and secrete an alkaline fluid which protects the lining of the duodenum.

(3) *Lymph glands.* These are found chiefly in the ileum, where there are solitary lymph glands called nodules, and groups of these glands called Peyer's patches.

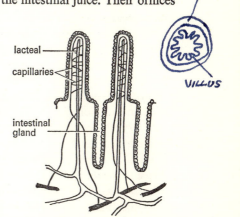

lacteal
capillaries
intestinal gland

MUSCULAR WALL
VILLUS

Fig. 76. Two villi of the small intestine.

Digestion in the small intestine

When the acid chyme enters the duodenum it acts on the mucous membrane and causes it to produce a hormone called secretin. This pours into the blood and passes via the blood to the pancreas, stimulating the latter to secrete its juice.

The acid chyme at the same time stimulates the intestinal glands to secrete their juice, and it also stimulates some afferent nerve endings which bring about a reflex emptying of the gall bladder, and bile is poured into the duodenum. The contraction of the gall bladder is also brought about by a hormone, cholecystokinin which, like secretin, is produced when certain foods act on the mucous membrane of the duodenum.

CHYME STIMULATES INTESTINAL GLANDS

Bile is produced in the liver and stored in the gall bladder. In man 17–35 ounces are secreted by the liver daily. After the gall bladder is removed at operations the bile is secreted continuously into the duodenum.

Bile is a yellowish-green alkaline fluid consisting of 86 per cent water and 14 per cent solid matter in solution. The solids include mucin, bile salts, and bile pigments (bilirubin and biliverdin). Bile contains no enzymes.

Bile pigments are purely excretory substances, being produced from broken-down red blood corpuscles. In the alimentary canal they are converted by bacteria into stercobilin, the colouring matter of the faeces.

Bile salts are important in digestion, for by lowering the surface tension between the fat and the water they help to emulsify fats, and a much greater surface of the fat is exposed to the action of the fat-splitting enzyme in the pancreatic juice.

Bile also acts as a weak antiseptic, and as a mild aperient, the mucus contained in it lubricating the contents of the duodenum.

(HORMONE SECRETIN)

Pancreatic juice. This is a colourless alkaline fluid, containing some solids in solution, including sodium bicarbonate and three enzymes, lipase, trypsinogen and pancreatic diastase (amylase).

Lipase acts on fats converting them into fatty acids and glycerol. Lipase is soluble in water, but insoluble in fat so that it can only act on the surface of the fat globules.

Trypsinogen is an inactive pro-enzyme. When it meets the enterokinase in the intestinal juice it is converted into trypsin. Trypsin can convert any protein not changed in the stomach into peptones, and it can also convert peptones into amino acids.

LIPASE — FATS.
TRYPSIN — PROTEIN.
AMYLASE — STARCH

AMYLASE

Diastase converts any starch not acted upon in the mouth into maltose. It is capable of acting upon uncooked starch.

Intestinal juice contains the following enzymes: erepsin, a mixture of several specific enzymes which convert peptones to amino acids; maltase, which converts maltose to glucose; lactase, which converts lactose to glucose and galactose; and invertase, which converts sucrose to a mixture of glucose and fructose.

In health, digestion is completed and the products of digestion are absorbed in the small intestine. The undigested food passes on into the large intestine.

Absorption

The fatty acids and glycerol pass into the columnar epithelial cells on the outside of the villi. There they unite to form minute droplets of fat which pass through the submucous coat into the lymph in the lacteal. The fat is then taken away by the lymph, which enters the blood in the internal jugular veins (see page 95).

The amino acids, glucose, fructose and galactose pass through into the blood in the capillaries inside the villi, though in the epithelial cells some of the amino acids may be converted into albumen and globulin, which form the blood proteins. The fate of the absorbed foods will be discussed in the next chapter.

The walls of the small intestine show different movements which help the organ to carry out its functions. The food is passed along the small intestine from time to time by peristalsis; it is mixed well by the pendulum or swaying movements of the organ; segmental movements, in which about every 8 seconds alternate portions of the wall are constricted, help the mixing of the contents, and also absorption by allowing fresh portions of the contents to come in contact with the mucous membrane.

When the wall of the small intestine is irritated peristalsis becomes very violent and the liquid contents are rushed on, and so the person suffers from diarrhoea.

Large intestine

The small intestine joins the large intestine at the ileo-caecal valve. The large intestine is about 5 ft long and extends from the ileum to the anus. It is divided into the caecum with the vermiform appendix, the colon, rectum and anal canal.

The caecum is the first part. It is a large sac with a blind lower end from which the appendix juts. Although the appendix is important in vegetable-eating animals for the digestion of cellulose, in man it has lost this function and if it becomes inflamed it has to be removed (appendicitis).

Above, the caecum is continuous with the colon; the first part or the ascending colon runs up the right lumbar region of the abdomen. Just below the liver the colon turns across the front of the abdomen at the hepatic flexure, and forms the transverse colon between the epigastric and umbilical regions of the abdomen. Beneath the spleen it turns down, forming the splenic flexure, and becomes the descending colon in the left lumbar region of the abdomen. In the left iliac region it bends, forming the sigmoid flexure, and enters the pelvic cavity to become the rectum, a tube about 5 in long.

The last $1\frac{1}{2}$ in of the alimentary canal is known as the anal canal. This opens to the exterior at the anus, an orifice guarded by sphincter muscles.

The wall of the large intestine has the same four coats as the wall of the small intestine, but its mucous membrane has no villi. The test-tube shaped glands of the mucous membrane are long and narrow, and secrete mucin.

Functions of the large intestine. The undigested food passes in a liquid state from the small intestine into the large intestine. In this organ water is absorbed into the blood, so that as the contents, now called the faeces, pass along by peristalsis they become more solid. The mucin secreted keeps them soft and well lubricated.

The walls of the large intestine excrete some unwanted mineral salts, e.g. iron, and calcium and magnesium phosphate, into the faeces.

Peristalsis is very slow in the large intestine, the faeces taking about 16 hours to reach the sigmoid flexure, where they remain for some time.

Defaecation is the act of passing the faeces to the exterior, and the desire to carry this out should take place at the same time each day. In a healthy person this is brought about by the gastrocolic reflex. The food passed into the stomach at breakfast stimulates a reflex peristalsis in the small intestine and later the colon. The undigested food passes on and the faeces in the sigmoid flexure pass into the rectum. The pressure set up there brings about a reflex, which results in the relaxation of the sphincter muscles of the anus, accompanied by the contraction of the walls of the sigmoid flexure and rectum, and the faeces are expelled. Though this is a reflex action, it is under control of the will and is helped by the voluntary contraction of the muscles of the pelvic floor and abdominal wall.

Faeces

The faeces consist of a solid or semi-solid mass containing 65–75 per cent of water. They are coloured by a pigment, stercobilin, formed from the bile pigments. The solid matter includes mucin, iron, phosphates of calcium and magnesium, shed epithelial cells from the intestinal wall, dead and living bacteria, undecomposed cellulose if present in the diet and nitrogenous matter derived from bacterial decomposition of undigested protein in the large intestine. The offensive odour is due to indole and skatole, which are derivatives of the amino acid tryptophan.

PRACTICAL

1 *Experiment with ptyalin*
   (*a*) Make a 'starch solution'. Add about 100 ml of water to a small table-spoonful of starch in a beaker. Heat to boiling-point and boil for two minutes, stirring all the time. Allow to cool.
   (*b*) Make a 'ptyalin solution'. Rinse the mouth well with warm water and transfer the water to a beaker.
   (*c*) Put 5 ml of the 'starch solution' into each of three test-tubes, A, B and C.
   (*d*) To test-tube A add 5 ml of 'ptyalin solution'.
   (*e*) To test-tube B add 5 ml of boiled 'ptyalin solution'.
   (*f*) Shake and place test-tubes in water-bath at about body heat (35–40°C).
   Remove drops of liquid from test-tubes every half-minute and add them to drops of iodine solution on a porcelain slab. Test samples of the solutions for reducing sugar. Note results carefully and give explanations.
2 *Experiment with pepsin* (this can be obtained in powdered form)
   (*a*) Add some chopped egg-white and 5 ml of water to four test-tubes A, B, C and D.
   (*b*) To test-tube A add four or five drops of dilute hydrochloric acid and a few grains of pepsin.
   (*c*) To test-tube B add a few grains of pepsin.
   (*d*) To test-tube C add four or five drops of dilute hydrochloric acid.
   (*e*) Shake the four test-tubes and place in water-bath at body temperature.
   After one hour, extract some liquid from each test-tube and test for protein with biuret test. Note results carefully and give explanations.
3 *Experiment with rennin*
   Add 3 ml of rennet (commercial rennin) to milk in a test-tube. Place this test-tube and another containing only milk (after shaking them well) in a water-bath at body temperature. Note result.
4 *Effect of salts on oil and water*
   Add 5 ml of olive oil and about 10 ml of water to two test-tubes A and B.

To test-tube A add a pinch of sodium bicarbonate. Shake both test-tubes and replace in test-tube rack. Note results.

5 *Experiment with lipase* (this can be obtained in powdered form)

Take two test-tubes A and B. To test-tube A add water, olive oil, and lipase. To test-tube B add water and olive oil, but no lipase. Shake well and leave the tubes to stand. Note the result.

NUTRITION   FIVE STAGES.

(1) INGESTION
(2) DIGESTION
(3) ABSORPTION
(4) ASSIMILATION
(5) EXCRETION.

AMINO ACIDS = PROTOPLASM
CYTOPLASM OF TISSUE CELLS ASSIMILATE
A/M. ACIDS BUILD THEM INTO PROTOPLASM
ENERGY PROD/. DURING TISSUE RESPIRATION
USED TO DO THIS (ENERGY REQUIRED FOR GROWTH)

SURPLUS AMINO ACIDS — LIVER — UREA
          ⤷ CANNOT BE STORED IN BODY.

FATTY ACIDS BURNT UP BY OXYGEN
DURING TISSUE RESPIRATION TO PROVIDE
                              ENERGY

13 Metabolism, the liver and pancreas

HEPATIC PORTAL VEIN — TO LIVER (FROM HEPATIC VEIN

LIVER ACT AS SUGAR REGULATOR

GLYCOGEN STORED IN LIVER

↳ RECONVERTED — GLUCOSE BY ADRENALIN

The amino acids, glucose and fructose, are absorbed into the blood in the capillaries in the walls of the small intestine. These capillaries join up to form the hepatic portal vein, which takes the blood with its contained foodstuffs to the liver.

Before discussing what happens to the various foodstuffs, the structure of the liver will be described.

The liver

The liver is the largest gland in the body, weighing about $3\frac{1}{2}$ lb. It is wedge shaped, the base of the wedge being on the right side. It lies in the dome of the diaphragm, and is thus protected by the lower part of the thoracic wall.

During respiration, when the diaphragm descends, the liver is compressed, and this rhythmic compression aids the circulation of blood through the organ. For this reason exercise is important for the proper functioning of the liver and the prevention of liver congestion. A congested liver is swollen, and the swelling can be felt below the last pair of ribs.

The liver is covered with peritoneum and is attached to the diaphragm by the peritoneal ligaments, and to the stomach and duodenum by a fold of peritoneum called the lesser omentum. This omentum is attached at a cleft, which divides the liver into the right and left lobes.

The gall bladder, which is a pear-shaped muscular bag, lies on the under surface of the right lobe of the liver. It is about $3\frac{1}{2}$ in long and is divided into the fundus, body and neck. Its wall has three coats, an outer serous coat, a middle muscular coat, and an inner mucous coat.

A duct called the cystic duct leads from the gall bladder and meets the hepatic duct, formed by the union of the two bile ducts, one from each lobe of the liver. The cystic duct and hepatic duct join to form the

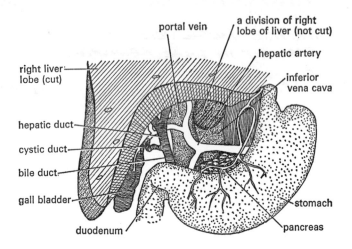

Fig. 77. Diagram to show relative positions of liver, gall bladder, and stomach.

common bile duct, which passes out of the liver in the lesser omentum and passes to the duodenum.

The hepatic artery and the hepatic portal vein enter, and the hepatic

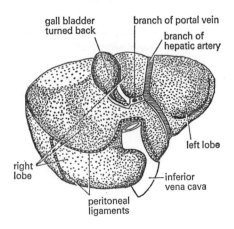

Fig. 78. Under surface of the liver.

vein leaves alongside the bile duct in the great omentum. Inside the liver these vessels divide into two, so supplying both lobes of the liver.

On the under surface of the liver the right lobe is further subdivided into three small lobes.

On microscopic examination of a liver section it is seen to be made up

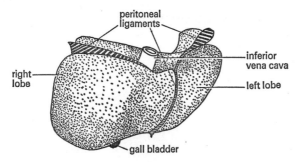

Fig. 79. Upper and front view of the liver.

of many-sided lobules about the size of a small pin-head. A hepatic lobule, or liver lobule, is made up of liver cells which are roughly cubical in shape and about $\frac{1}{1000}$ in in diameter. Each cell has a large nucleus and granular cytoplasm. The cells generally contain glycogen granules and sometimes fat globules. It can be proved by chemical tests that they contain iron.

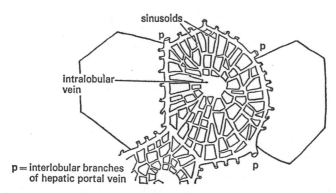

Fig. 80. Hepatic lobules (highly magnified).

The cells are arranged in columns radiating from the centre of the lobules. They are supported by reticular tissue, which forms a network and penetrates from the connective tissue surrounding each lobule. There is a very thin layer of connective tissue underneath the peritoneum covering the liver.

Branches of the hepatic artery and the hepatic portal vein run through the organ, surrounded by connective tissue and accompanied by bile ducts. Smaller branches run in the connective tissue between the lobules, and are known as interlobular branches. These break up into blood sinuses, known as hepatic sinusoids, which pass between the cells to the

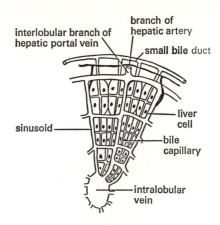

Fig. 81. Section of hepatic lobule in more detail.

centre of the lobules. A sinusoid is wider and more irregular than a capillary, its wall being formed partly by flattened endothelial cells and partly by branched cells of the reticulo-endothelial system. Every liver cell is in contact with a sinusoid. The sinusoids formed by the branching of the hepatic artery, and the sinusoids formed by the branching of the hepatic portal vein, join together, so that there is no distinction between the two sets. In the centre of the lobule the sinusoids unite to form the intralobular vein. This unites with other intralobular veins to form a larger vein, and finally a large vein from the right lobe unites with a large vein from the left lobe to form the hepatic vein, which takes the blood away from the liver.

The liver cells discharge bile into very small canaliculi. These communicate with the bile capillaries, which form a network in the lobules in between the cells. The bile capillaries join up to form the small bile ducts on the outside of the lobules. The small bile ducts join up to form larger ducts, which in their turn join until finally a duct known as the hepatic duct leaves the liver.

The fate of absorbed carbohydrates and fats. The glucose and fructose pass to the liver in the hepatic portal vein and are converted by means of a liver enzyme into glycogen, or 'animal starch' which is stored in the liver cells. The liver allows some glucose to remain in the blood, so that the sugar concentration of blood plasma is kept at approximately 0·1 per cent. As the blood circulates, the tissues which require a source of energy are continually removing the sugar, but when the blood sugar concentration goes down the liver reconverts some of its stored glycogen into glucose.

When the muscles take out the glucose from the blood, they do not utilize it immediately, but convert it into glycogen, which they store until it is needed.

The fat absorbed into the lacteals passes in the lymph and enters the blood at the left internal jugular vein. It then passes round the body and is laid down in certain fat depots, e.g. under the skin and round internal organs.

When more carbohydrate than is taken in is required by the body, the liver soon becomes empty of glycogen. When this happens the fat from the fat depots passes in the blood to the liver, where it is made suitable for utilization by the tissues.

If on the other hand excess carbohydrate is taken in, this can be converted into fat and stored.

The fate of the absorbed amino acids. The absorbed amino acids pass in the hepatic portal vein to the liver. They are not stored there, but pass round the body in the blood, and the cells as they require them remove the suitable amino acids for building up their protoplasm. The amino acids not used in this way pass back in the blood to the liver. There they are de-aminated, that is the nitrogen part of the molecule is removed and converted into urea. This urea, called exogenous urea since it is derived solely from the food, passes in the blood to the kidney, where it is excreted. The remaining part of the amino acid molecule can then be converted into carbohydrate or fat and used as a source of energy.

Functions of the liver

The functions of the liver can now be summarized:

(1) The liver stores excess carbohydrates as glycogen and regulates the amount of glucose in the blood.

(2) The liver prepares fat for utilization as a source of energy.

(3) The liver de-aminates amino acids not needed for building up new protoplasm. From some amino acids it is able to manufacture fibrinogen and probably albumin proteins of blood.

(4) During the 'wear and tear' of the body a certain amount of protoplasm is broken down and forms nitrogenous waste. This passes in the blood to the liver, where it is converted into urea, creatinine or uric acid, endogenous nitrogenous waste derived from tissue waste (see page 146). This waste then passes in the blood to the kidneys, where it is excreted.

(5) The liver produces and secretes bile. Normally bile is stored in the gall bladder, which empties its contents into the duodenum from time to time (see page 126).

(6) The storage of the anti-anaemic principle, now known as Vitamin B_{12} (see page 124). The liver discharges this into the blood in which it passes to the bone marrow where it has some effect on the formation of new red corpuscles.

(7) Storage of iron.

(8) The liver is a large organ and does a great deal of work. Consequently it produces much heat energy, which is important in keeping up the body temperature. For this reason, when the liver is not working properly, a person feels chilly.

The pancreas

The pancreas is a gland which produces pancreatic juice. This passes down the pancreatic duct into the duodenum, where it helps digestion. The pancreas also secretes a substance, insulin, directly into the blood. Before discussing the importance of insulin the structure of the pancreas will be described.

The pancreas is a compound racemose gland, very similar to the salivary glands. It is a soft gland, about 7 in long, extending from the duodenum to the spleen. It is made up of a head, neck, body and tail. The head is encircled by the duodenum, the neck projects from the head on the anterior side and unites it with the body. The body lies behind the stomach to the left side and joins the tail of the pancreas, which just touches the hilum of the spleen (see Fig. 75).

Microscopic examination of a section of the pancreas shows secretory alveoli. Each alveolus consists of a central duct, lined by a single layer of polyhedral cells on a basement membrane. The alveoli are collected together by connective tissue into lobules.

It is the alveoli cells which produce the pancreatic juice. They pass it into the central ducts, which join up into larger ducts. These larger ducts join up until the pancreatic duct leaves the pancreas.

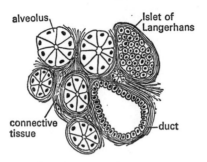

Fig. 82. Section of pancreas.

Scattered between the lobules of the pancreas are clumps of cells called the islets of Langerhans. The cells forming these islets have no ducts, the insulin which they produce passes directly into the blood. The pancreas, like all secretory glands, is highly vascular.

Functions of the pancreas

The pancreas has two important functions:

(1) The secretion of pancreatic juice (see page 126).

(2) The secretion of insulin. The insulin circulates round the body in the blood, and is important in carbohydrate metabolism. Without insulin no glycogen can be stored in the liver and muscles, and glucose cannot be oxidized to produce energy.

Thus when the islets of Langerhans are diseased no insulin is formed. The excess carbohydrate absorbed from the small intestine cannot be stored in the liver, and the blood sugar concentration goes up. This condition is known as hyperglycaemia. If the sugar concentration rises above about 0·18 per cent (the kidney threshold) sugar is excreted in the urine.

In normal individuals there is no sugar in the urine unless an excessively large dose of sugar (far more than 100 g) is taken by the mouth. Even then it may not be found, for in the majority of individuals the liver can store up excess glucose almost as fast as it is absorbed by the blood, and the concentration never rises above the renal threshold (see page 147). Thirty minutes after a large glucose meal the blood sugar may rise to 0·18 per cent, but at the end of two hours the blood sugar in the normal individual is back to the fasting level.

Without insulin, the body cannot get its energy by oxidizing sugar, and it strives to rectify this by converting fats and amino acids into glucose. This, when formed, cannot be stored or utilized, and is excreted by the kidneys. The body as a result wastes away. At the same time the unconverted fat cannot be properly oxidized without some sugar oxidation, and poisonous substances called ketones are produced. These appear in the urine and give a peculiar odour to the breath.

This disease produced by lack of insulin in the blood and characterized by the symptoms just described is called diabetes mellitus, and if lack of insulin is due to the improper working of the pancreas, the disease is known as pancreatic diabetes.

Pancreatic diabetes can be controlled by insulin injections. If too much insulin is injected the blood sugar concentration goes down, a condition known as hypoglycaemia. At the same time excess carbohydrate is stored and utilized. The condition may be relieved by administering large doses of glucose by the mouth.

It is now thought that the blood sugar concentration is controlled by a centre in the fore-brain.

PRACTICAL

1 Examine a piece of mammalian liver under a magnifying-glass. Note the mottled appearance. The small areas seen are lobules. Cut open the liver and notice branches of hepatic vein and hepatic portal vein inside the organ.

2 Examine a section of liver under microscope.

3 Examine a section of the pancreas under microscope.

14 Excretion and the regulation of body fluids

During the activities of the living cells certain waste substances are produced. If these accumulate in the body they become harmful. They are, however, passed to the exterior by certain organs, and this process is known as excretion.

The chief excretory organs in man are:

(1) The lungs, which excrete carbon dioxide and some water, the waste products of respiration.

(2) The skin, which gives out water and some mineral salts in the sweat.

(3) The urinary organs, consisting of two kidneys, two ureters, the bladder and urethra, which excrete nitrogenous waste, certain mineral salts and water.

The skin

The skin is the outer covering of the body, and it consists of two layers, an outer epidermis and an inner dermis, which is attached to underlying structures by loose strands of connective tissue.

The dermis is made up of fibrous and elastic connective tissue; on its outer surface these fibres rise up into folds or papillae, which appear through the epidermis and form the finger-prints.

There are nerve endings, sense organs (see Chapter 17), blood vessels and lymphatic vessels in the connective tissue. There are also sweat glands consisting of coiled tubes lined with secreting epithelium and surrounded by a network of capillaries. The ducts of these glands pass through the epidermis to open at pores on the surface of the skin. If the skin of the palm is examined with a magnifying-glass these pores are seen to be arranged in rows on the ridges. On the palm of the hand there are at least 3,500 sweat glands per square inch. They are more numerous under the arms and between the legs.

Underneath the dermis and lying on top of the deeper structures is a layer of fat, the subcutaneous fat, which gives roundness and softness to the body. It acts as a buffer and protects the body from mechanical injury.

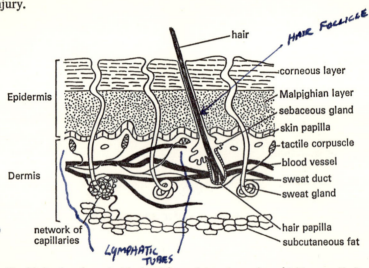

Fig. 83. Section through skin (at right angles to the surface; highly magnified).

The epidermis is moulded over the papillae of the dermis. The layer next the dermis consists of columnar cells with a definite nucleus and pigment granules. It is these granules which are dark in coloured skins and which change colour when the skin becomes sunburnt. This layer is called the Malpighian layer. Its cells are constantly dividing, and the new cells become pushed out towards the surface. At first they are almost round and contain granules and protoplasm, but they eventually die and the cells become flat and horny and form the corneous layer of horny stratified epithelium on the outside of the skin. The outermost cells are continually rubbed off, but are replaced by more cells being pushed outwards from below. The thickness of the corneous layer varies in different parts of the body, being thickest on the soles of the feet and the palms of

MELANIN = DARK SKIN

the hands. There are no blood vessels in the epidermis, the living cells being nourished by the fluid which exudes from the dermal blood vessels.

The hairs and nails are appendages of the skin, produced from the epidermis. In the development of a hair the innermost layer of the epidermal cells sinks down into the dermis to form the hair follicle in the base of which a hair papilla forms. This papilla consists of a mass of dermal tissue containing blood vessels and nerves covered with epidermal cells. These epidermal cells push out cells which die and form the hair. If a healthy hair drops out the papilla soon produces a new one. Baldness is due to decay of the hair papilla.

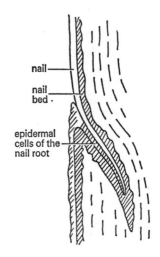

nail

nail bed

epidermal cells of the nail root

Fig. 84. Longitudinal section through the nail.

Sebaceous glands open into the hair follicle and also independently. These produce and secrete a fatty substance called sebum, which keeps the hair glossy and the skin smooth.

The colour of the hair depends on the pigment in the epidermal cells producing the hair. When air bubbles collect between the dead hair cells the hair goes grey.

Nails are formed in a similar way. The nail root at the base of the nail is covered with skin. The epidermal cells of the root cover a mass of dermal tissue and 'cut off' cells which are hard and horny and are pushed out to form the thick body of the nail – the part exposed. This body of the nail rests on a nail bed, which is ridged, very vascular and highly sensitive, containing a relatively large number of nerve endings.

Excretion by the skin

The sweat glands are continually giving out water containing small quantities of sodium chloride, organic matter and carbon dioxide. Normally the sweat evaporates as soon as it is formed, and is known as insensible sweat, but in hot and humid weather and after active exercise drops of water are seen on the skin.

Other functions of the skin

(1) *Protection.* The skin, being the outer layer of the body, protects the

underlying organs from mechanical injury (helped in this by the sub-cutaneous fat); changes in temperature; invasion by germs, etc.

(2) *Regulation of body temperature.* This will be dealt with in the next chapter.

(3) *Sensory.* In the dermis are tactile corpuscles, sense organs for perceiving stimuli of touch and temperature changes. These will be described in the chapter on the special senses.

The kidney

The urine containing the nitrogenous waste and unwanted mineral salts is produced in the kidneys, which remove these substances from the blood. The formation of urine is continuous, and the urine passes down

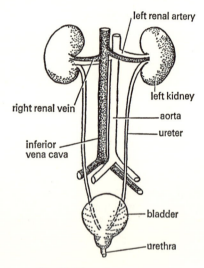

Fig. 85. The urinary organs.

the ureters to a pear-shaped sac, the bladder, situated in the pelvis. Here the urine collects. The bladder is emptied from time to time, the urine being expelled to the exterior through a tube called the urethra.

The kidneys, which are surrounded by a mass of fat, lie in the posterior part of the abdomen behind the peritoneum, one on either side of the vertebral column. The upper ends are level with the last thoracic vertebra and the lower ends with the third lumbar vertebra. The right kidney is slightly lower, leaving room for the right lobe of the liver, which extends farther down than the left. The kidneys are 3–4 in long and 1–1½ in wide, the left kidney being a little longer and narrower than the right.

A kidney has a characteristic shape, the convex border being directed away from, and the concave border directed towards the vertebral column. In the middle of the concave border is a deep vertical fissure, the hilum, where the renal artery and nerves enter, and the renal vein and ureter leave.

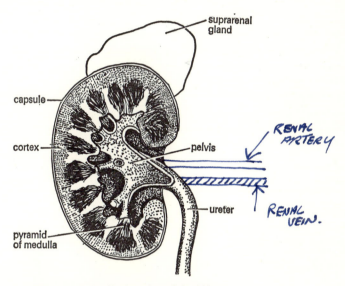

Fig. 86. Longitudinal section of the kidney.

Inside the kidney the ureter begins as a funnel-shaped cavity, the pelvis. This structure can be seen in a longitudinal section (see Fig. 86).

Around the pelvis is a lightish red medulla which has 12–16 conical shaped processes or pyramids projecting into the pelvis. Surrounding the medulla is the darker red cortex, and the whole kidney is surrounded by a capsule of fibrous tissue which forms a smooth covering.

The kidney is made up of a large number of minute tubules bound together by connective tissue. These can be seen with a microscope.

Each tubule begins in the cortex as a capsule (Bowman's capsule), which is similar to a cup with hollow walls. The capsule leads into a convoluted tubule which is followed by a loop in the medulla known as the loop of Henle. This leads into a second convoluted tubule in the cortex and then many tubules unite to form a collecting duct. This duct passes through the cortex and the medulla, and ends at the apex of one of the pyramids.

The walls of the kidney tubules consist of a layer of cells one cell thick. The cells have distinct nuclei.

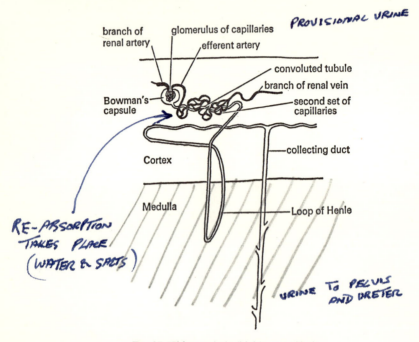

Fig. 87. Kidney tubule (highly magnified).

Blood supply. A renal artery carries blood to the kidney. It divides up within the organ and a small arteriole takes blood to each tubule. The arteriole breaks up into capillaries, forming the glomerulus inside the capsule cup. The capillary walls are fused with the very thin walls of the capsule. The capillaries join up to form a small artery, the efferent artery, which carries the blood away from the capsule. This vessel breaks up into a second set of capillaries around the convoluted tubules and these capillaries join up to form a small vein which carries the blood away from the tubule. Then all the small veins unite to form the renal vein, which carries blood away from the kidney.

Formation of urine. The concentration of certain substances found in the blood and normal urine are compared in the following table:

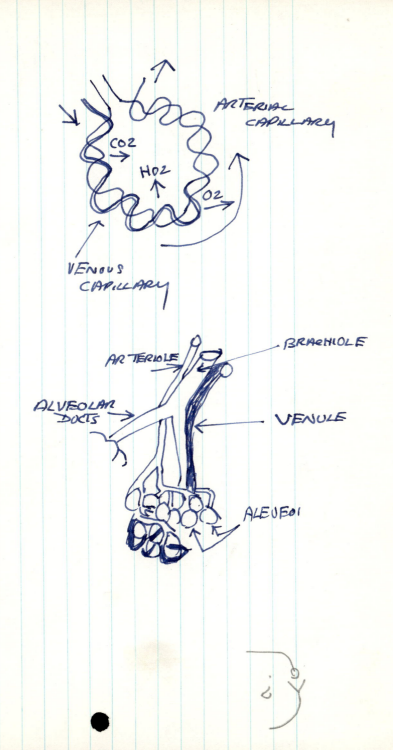

ARTERIAL CAPILLARY

CO2

HO2

O2

VENOUS CAPILLARY

ARTERIOLE

BRACHIOLE

ALVEOLAR DUCTS

VENULE

ALEVEOI

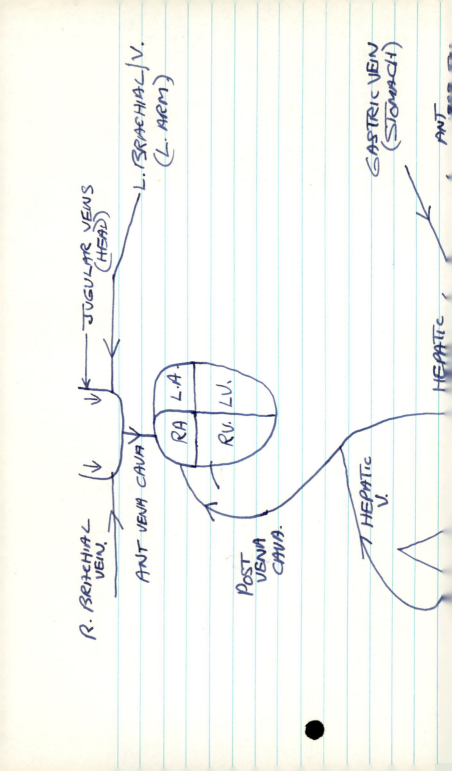

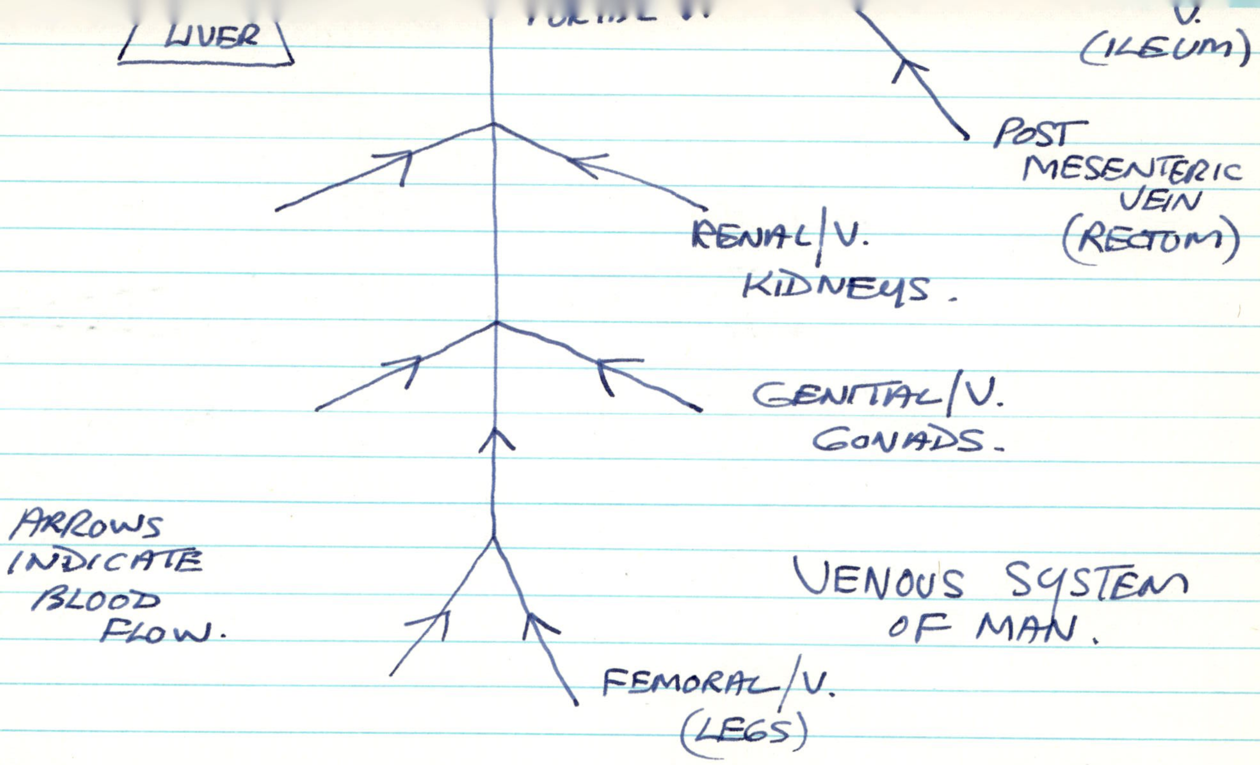

V.
(ILEUM)

POST
MESENTERIC
VEIN
(RECTUM)

RENAL/V.
KIDNEYS.

GENITAL/V.
GONADS.

VENOUS SYSTEM
OF MAN.

FEMORAL/V.
(LEGS)

LIVER

ARROWS
INDICATE
BLOOD
FLOW.

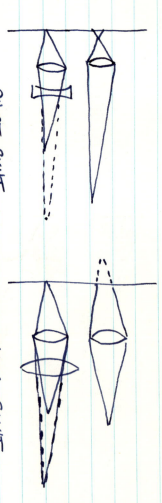

SHORT SIGHT

LONG SIGHT

	% *Blood*	% *Urine*
Protein	7–8	0
Glucose	0·1	0
Salt (sodium chloride)	0·7	0–0·4
Urea	0·03	2
Creatinine	0·001	0·075
Uric acid	0·004	0·059

The kidney does not act as a simple filter. The membrane forming the Bowman's capsule is semi-permeable, i.e. permeable to simple substances such as sugar and salt in solution, but impermeable to protein, and the

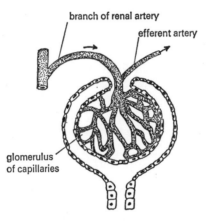

Fig. 88. A Bowman's capsule.

plasma without its proteins filters through this membrane into the tubule. This theory is supported by experiments performed in the frog. In this animal the glomeruli lie near the surface and can be seen with a dissecting microscope. If a pipette is placed in the space between the walls of the capsule and some liquid extracted it is found to be blood plasma without its protein. This liquid is not urine; something must happen to alter the concentration of the various solutes before the liquid reaches the ureter. It is thought that sodium chloride and glucose are removed from the liquid in the first convoluted tubule and the loop of Henle and that nitrogenous waste is secreted into the second convoluted tubule. To carry out selective absorption and separate water from its solutes a great deal of energy is required. The fact that the kidney of the frog consumes much oxygen and therefore must be using up energy is evidence that the process of urine formation is not one of passive filtration.

There is no direct evidence for this process in the kidney of man, but it is assumed to be similar. Man excretes about 1½ litres of urine a day, but probably as much as 100 litres of liquid filter through the capsules, a great amount being reabsorbed into the blood together with the dissolved sugar and salt.

Normal urine. Normal urine is a clear pale amber-coloured liquid with an aromatic smell. It is slightly acid to litmus and has a specific gravity of 1·015–1·025.

It consists on an average of about 96 per cent water and 4 per cent solids in solution. The chief solids are:

urea	about 50 per cent of the total solids.
uric acid	in the form of urates.
creatinine	
ammonia	in the form of ammonium salts.
sodium chloride	about 60 per cent of the inorganic constituents.
phosphoric acid	
sulphuric acid	
magnesium	in the form of inorganic salts.
calcium	
potassium	

Urea, uric acid and creatinine all contain nitrogen. Urea forms about 90 per cent of the nitrogenous material excreted. It is formed in the liver chiefly from the ammonia, which has been broken off from the unwanted amino acids of the diet, partly from the nitrogenous waste produced during the 'wear and tear' of the cells. The amount of urea excreted varies greatly with the amount of protein in the diet.

Creatinine is produced from the breakdown of the protoplasm of striated muscles. The amount excreted increases during great muscular activity, and in certain wasting diseases, including the later stages of starvation. It is not affected by the protein of the diet.

Uric acid is produced from the nuclear material either of the actual body cells or of the food.

The amount of ammonia excreted varies with the acidity of the other urinary constituents. Normal urine is slightly acid, though the reaction does vary with the food, sometimes becoming alkaline. If a number of very acid salts are formed some amino acids are converted into ammonia which neutralizes these acids.

In health the concentration of the various salts of the blood plasma is practically constant. This regulation is brought about by the kidney. The small amounts of salts which are required by the body are taken in with the food and any excess is excreted through the kidneys. When the salt intake is small or when salts are disposed of in other ways their concentration in the urine rapidly decreases. For example, during hot weather when a large amount of sodium chloride is lost in the sweat, and the water supply of the body is kept up by drinking salt-free liquid, the urine excreted though concentrated has a low salt content. Men engaged in heavy manual work which entails sweating require much salt, and if this salt is not supplied, a condition in which the muscles go into spasms, called miner's cramp, is produced. If the food contains a large amount of calcium and magnesium the phosphates in the urine diminish, as they are excreted in the faeces in the form of the rather insoluble calcium and magnesium phosphates. When the amount of calcium or magnesium to be excreted is small the phosphates are excreted in the urine in the form of the very soluble sodium phosphates.

Abnormal constituents of urine. In disease the urine may contain some normal constituents in excess and some substances which are not normally present.

Glucose in the urine has already been mentioned (see Chapter 13). It is excreted whenever the glucose in the blood rises above 0·18 per cent (the kidney threshold). This occurs in diabetes (see page 138). In severe cases of this disease the urine also contains ketone bodies and fatty acids.

Proteins, e.g. serum albumin and serum globulin. Proteins appear in the urine when the kidney is damaged in any way, e.g. in damage due to inflammation of the kidney or nephritis.

Amount of urine and diuresis. Normally the kidney excretes about $1\frac{1}{2}$ litres of urine a day. In hot weather, when more water is excreted by the sweat, a smaller volume of more concentrated urine is given out. Such urine is darker in colour.

Diuresis or increased flow of urine is produced under certain circumstances, e.g.

(1) When the blood pressure in the renal artery goes up. This is caused by a general rise in blood pressure, or by a constriction of the small veins, or relaxation of the arterioles.

(2) When the blood plasma becomes diluted. This is caused by an excessive ingestion of water or salt-free liquid. The plasma proteins become diluted and do not exert such a high osmotic pressure on the rest of the plasma filtering through the Bowman's capsule and more liquid filters.

(3) When the blood sugar concentration rises above the renal threshold.

(4) When fluid loss by the skin is inhibited, for example, when the skin is exposed to cold.

(5) When certain drugs known as diuretics, e.g. caffeine, are present in the blood.

(6) When the amount of pituitrin in the blood decreases.

Pituitrin is a hormone secreted by the posterior lobe of the pituitary gland. If a person takes in a large quanity of water and pituitary extract is then injected, no excretion of urine takes place for at least 24 hours. If the gland is diseased the urine is formed in great quantities. It is probable that in health pituitrin is continually being secreted into the blood, and the water balance of the body is controlled. The disease caused by defective pituitary action and characterized by a large flow of urine is known as diabetes insipidus. Antidiuresis or decreased urine flow is caused by:

(i) Lowering of the blood pressure in the renal artery.

(ii) Loss of water in some other way, e.g. in the sweat.

(iii) Excess pituitrin in the blood due to overactivity of the posterior lobe of the pituitary gland.

Functions of the kidneys

The kidneys, in forming and giving out urine, carry out the following functions:

(1) They excrete waste products of metabolism and digestion.

(2) They control the concentration of the individual salts of the blood plasma.

(3) They control the total solid concentration, i.e. the osmotic pressure of the blood.

(4) They control the reaction of the blood plasma, that is they maintain the alkalinity of the blood and other body fluids.

(5) They remove any substances produced in any abnormal metabolism of the body.

The urine formed in the kidney passes down the ureter to the bladder. A ureter is a narrow tube about 14–16 in long. Its wall has three coats, an outer fibrous one, a middle muscular one and an inner mucous coat. The two ureters enter the bladder obliquely, and this prevents any regurgitation of urine.

The bladder is a pear-shaped organ, its size varying with the amount of urine it contains. It lies in the pelvis behind the symphysis, and its base is fixed while its upper part or fundus is free. The fundus rises up and extends into the abdominal cavity, when the organ becomes distended with urine.

The bladder has four coats:

(1) An inner mucous coat which is continuous with the mucous coat of the ureters and urethra. This coat is folded in the empty bladder. The folds disappear when the bladder is distended.

(2) A submucous coat of areolar tissue connecting together the muscular and mucous coats.

(3) A muscular coat consisting of three layers of unstriped muscle fibres.

(4) A serous coat. This is derived from the peritoneum and covers only the upper parts.

Around the neck of the bladder are muscle fibres which act as a sphincter known as the internal sphincter; when contracted they keep the orifice closed.

The urethra, which is a canal $1\frac{1}{2}$ in long in the female, 8–9 in in the male, leads from the neck of the bladder and opens to the exterior at the urinary meatus. Its walls contain a layer of circular muscles which form the urethral sphincter or external sphincter which after infancy is under voluntary control.

Micturition

This is the act of passing urine from the bladder to the exterior. The mechanism can be understood better if the nerve supply to the bladder and sphincter is studied.

These organs receive a nerve supply from two sets of nerves. Motor fibres pass:

(1) In the pelvic nerves. Stimulation of these causes the bladder to contract and internal sphincter to relax.

(2) In the sympathetic system (fibres pass from the hypogastric plexus). Stimulation of these causes inhibition of the bladder muscles and contraction of the internal sphincter.

Sensory fibres pass from the bladder in the pelvic and hypogastric nerves to the spinal cord.

The urine is continually collecting in the bladder and the muscular walls become distended. They attempt to contract and expel the urine, but normally the internal sphincter muscles are in a state of tonic contraction. Eventually, distension stimulates the sensory nerve endings in the bladder wall, impulses pass up to the spinal cord and then impulses pass down the pelvic nerve to the bladder and internal sphincter, tending to make the former contract and the latter relax. Micturition is, however, under the control of the higher centres, which can keep the external sphincter contracted, only allowing expulsion of urine at certain times.

Under normal circumstances the desire to micturate is felt when the pressure has risen in the bladder to 150 mm water. If the urine accumulates slowly the distension of the muscular wall goes on slowly and much urine may collect before the critical pressure is reached. If the urine accumulates more rapidly the critical pressure is reached when a relatively small volume of urine has collected.

If the spinal cord is damaged above the centre of the micturition reflex in the lumbar region of the spinal cord, micturition is no longer under the control of the will and it becomes a reflex action. If the micturition lumbo-spinal centre itself is damaged the bladder will no longer be under nervous control and will behave as plain muscle, expelling part of its contents when it becomes distended. It cannot empty itself completely, and some urine will always remain in.

PRACTICAL

The skin

1 Mount a fragment of the corneous layer from the palm of the hand, tease in water and examine under the microscope. Note the cells.

2 Examine slides of sections of the skin and scalp under the microscope.

3 Examine a hair under the microscope. Note the core of cells covered with overlapping cells.

The urinary organs and urine

1 Examine the position of these organs in a rabbit. Note blood supply.

2 Examine a sheep's kidney. Cut longitudinally and examine macroscopic internal structure.

3 Examine slides of L.S. and T.S. of the kidney.

4 Examine a slide of T.S. of ureter.

5 Examine a slide of the muscular coat of the bladder.

6 Urine tests. Details for these must be obtained from a practical physiology book.

15 Body heat and the regulation of body temperature

Man, like birds and other mammals, is a warm-blooded animal. That is, his internal temperature does not vary with his surroundings, but remains practically constant at 36·9°C (98·4°F). A man at the North Pole has the same temperature as a man in the Tropics. If it varies much he is ill, for it means that the organs of the body are not functioning properly.

The body temperature is kept uniform in this way by a careful balance between heat production and heat loss.

Heat production

Heat is produced during internal respiration when the living cells are oxidizing food, the muscles and the liver being the most important furnaces. It has been shown that the blood leaving the liver is 1°F. warmer than the blood entering. The blood circulating round the body distributes the heat evenly.

Heat production is controlled by the nervous system. Variation in muscular tone of a muscle depends on the number of stimuli received from the central nervous system. This tone does not necessarily affect the position of the limb since it stimulates equally both extensor and flexor muscles.

In hot weather tone is low, and when sitting at rest muscles are flabby. Lassitude results, and a person has no inclination to perform muscular exercise. In cold weather muscle tone is increased, with the result that more heat energy is produced. When sitting at rest the muscles feel tense.

In cold weather a person is more active and by muscular movements produces more heat energy. At the same time more food is required to form a source of this extra energy.

When the body is cold, heat production may be increased by a twitching of the muscles or shivering. When this affects the jaw-closing muscles, chattering of the teeth results.

Heat loss

Heat may be lost from the body in several ways:
(1) By the faeces and urine.
(2) By the breath.
(3) By the skin.
The amount of heat lost by the faeces and urine remains approximately constant; the amount lost by the breath varies with the temperature and dryness of the air. In both these cases the amount lost cannot be controlled by the central nervous system.

The amount of heat loss through the skin varies, and depends on the amount of blood flowing through the blood vessels in the dermis, and the amount of sweat secreted.

When the temperature of the body rises, the blood vessels in the skin, which are normally slightly constricted, open out and blood flows to the skin. The skin becomes pink and warm and heat is lost to the outside air by conduction, radiation and convection.

On the other hand, when the temperature of the body falls, the blood vessels of the skin are constricted and the skin becomes cold, white, and is commonly known as 'gooseflesh'. Heat loss is reduced, as the heat must pass through the subcutaneous layer of fat before reaching the surface.

The controlling centre for the peripheral skin blood vessels lies in the medulla oblongata.

Sweat is continually being secreted from the sweat glands, but normally it evaporates as quickly as it is produced. In hot weather or during exercise the sweat is produced much more quickly, and may collect in drops of perspiration. Body heat is used up in evaporating this perspiration, and sweating cools the surface of the body. Damp clothes have a similar effect to sweating.

The sweat glands are under the control of the sympathetic system. Normally they are stimulated to secrete at the same time as the blood vessels in the skin are stimulated to dilate. As a result sweating accompanies flushing of the skin. During periods of fear impulses may pass down the sweat nerves and the vaso-motor nerves to the blood vessels and produce sweating, accompanied by constriction of the blood vessels. The condition is known as a 'cold sweat'.

A certain amount of heat loss is prevented by clothes. The material of which they are made entangles the air in its meshes, and so forms a bad conducting layer on the outside of the body. In other mammals heat loss is reduced by the outer covering of hair or fur.

The heat-regulating centre of the brain

The balance between heat loss and heat gain is adjusted by a heat-regulating centre, which lies somewhere in the forepart of the thalamus and affects centres in the medulla which control blood vessels and sweat glands. It is activated by the temperature of the blood passing over it.

Effect of hormones on body heat

An increase of heat production is brought about by the action of low temperatures on the nervous system. The secretion of hormones aids this nervous mechanism, e.g. thyroxin increases heat-producing activities and adrenalin reduces them by constricting the blood vessels.

Fever

When the body becomes infected with micro-organisms, fever or a rise of body temperature occurs. There are three stages of fever. During the rise of temperature heat production is increased by shivering, and heat loss prevented by the skin being dry. The skin becomes flushed and the pulse rate and breathing rate are both increased. During the second stage the temperature remains steady at a high level. When the body has overcome the infection the third stage sets in, muscle tone diminishes and heat loss is increased by sweating and the temperature then falls. The pulse and respiration return to their normal rate.

Heat stroke

In hot climates sometimes the sweat glands cease to function properly, or the air is so damp that little evaporation of perspiration can take place. Consequently when any muscular exercise is taken the temperature rises

F

and heat stroke is produced. It has harmful effects on the central nervous system, quickly producing headache, vomiting, dizzy fits and even convulsions.

Effect of climate on heat loss

A hot moist climate is unhealthy. The body is not cooled by sweating, since the evaporation of perspiration goes on very slowly in a damp atmosphere.

A hot dry climate is healthy. The dry air increases the evaporation of sweat from the skin, and so the temperature of the body is kept down. For comfort it is necessary for the water in the body to be replaced.

A cold moist climate is unhealthy. Moist air holds a great deal of heat and conducts it well, so that the heat loss from the body is relatively high.

A cold dry climate is healthy. Dry air holds little heat and conducts it badly, so that the heat loss from the body is relatively low.

PRACTICAL

1 Practise taking temperatures with a clinical thermometer:
   (a) under the armpit;
   (b) under the tongue, before and after exercise.

2 Place arm in bath of hot water. Note the immediate flushing of the skin.

3 Wet and dry bulb thermometer. Hang up two thermometers. Make sure that they both register the same temperature. Attach the end of a piece of cotton material about 3 or 4 in long round the bulb of one, and let the other end of the cotton hang in a beaker of water. Read the temperatures from time to time and explain the result.

16 The nervous system

The function of the nervous system is to enable the body to adapt itself to changing environment by responding to external stimuli. It co-ordinates the functioning of the various systems which in health work together in perfect co-ordination. The brain, the main organ of the nervous system, is the seat of consciousness, memory and reasoning power.

The nervous system is made up of:

(1) The spinal cord with its spinal nerves.
(2) The brain with its cranial nerves.
(3) The autonomic system (see page 168).
(4) The sense organs (see Chapter 17).

The spinal cord is a hollow tube which runs in the spinal canal of the vertebral column, and is therefore protected by bone. At the upper end it expands to form the brain, which is protected by a bony case, the skull. The spinal cord and brain together form the central nervous system (C.N.S.).

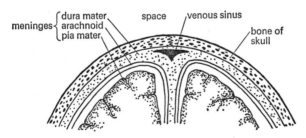

Fig. 89. Diagram of a section of the cerebrum to show meninges.

Between the bone and this central nervous system there are three membranes called meninges. They are the pia mater, the arachnoid and the dura mater.

The pia mater is adherent to the brain and the spinal cord, resting in between all the fissures, ridges and convolutions. It carries small blood vessels to and from the central nervous system.

The arachnoid membrane is a serous membrane which secretes cerebro-spinal fluid. The latter fills the space between the arachnoid and the pia mater, and also the cavity within the central nervous system.

The dura mater is a tough covering lining the inside of the skull and neural canal. It is part of this dura mater which forms the venous sinuses mentioned on page 83. The dura mater sends off septa, which partially divide up the cranial cavity.

Spinal cord and spinal nerves

The spinal cord which is nearly cylindrical is about 17 in long and ⅓ in in diameter. It passes down in the neural canal to the level of the upper border of the second lumbar vertebra, where it tapers into a conical extremity. The latter gives off a slender thread, the filium terminale which descends as far as the first segment of the coccyx.

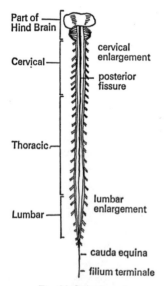

If the spinal cord is cut across an H of grey matter is seen around the small central canal. The grey matter is sur-rounded by white matter as seen in Fig. 91.

Thirty-one pairs of spinal nerves leave the spinal cord. These are:

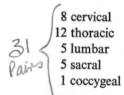

31 Pairs
{
8 cervical
12 thoracic
5 lumbar
5 sacral
1 coccygeal
}

Fig. 90. Spinal cord.

Each nerve leaves the spinal cord by two roots, which join after leaving the vertebral canal. The nerve roots of the lumbar and sacral nerves run almost vertically before leaving the spinal canal and form a cluster called the cauda equina owing to its resemblance to a horse's tail, around the filium terminale. After the union of the two roots a ramus communicans is given off, which contains fibres to the sympathetic system (see page 168).

A spinal nerve consists of both sensory and motor neurones; all the sensory neurones enter the spinal cord by the dorsal or posterior root and have their nerve cells in the dorsal ganglion; all the motor neurones leave the spinal cord by the ventral or anterior root and have their nerve cells in the anterior horn of grey matter of the spinal cord.

If the anterior roots of nerves supplying the legs, for example, are cut, the limbs lose their power of movement. If, on the other hand, the posterior roots are cut, the limbs lose their sense of feeling.

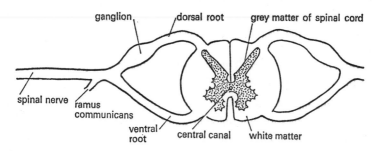

Fig. 91. Transverse section of spinal cord.

The spinal nerves divide into two branches, the anterior and the posterior primary rami.

All the posterior rami pass back to supply the muscles and skin of the back of their own region. In all regions except the thorax the anterior primary rami form nerve plexuses. In the thoracic region the rami circle round the thorax and supply the muscles (the intercostal muscles) and the skin between the corresponding pair of ribs.

The remaining primary rami form four plexuses:

(1) **The cervical plexus.** This is formed by the anterior rami of the first four cervical nerves joining together and forming a network which lies in the neck below the sterno-mastoid muscle. It sends out the phrenic nerves to the diaphragm and many branches to supply the neck muscles.

(2) **The brachial plexus.** This is formed by the anterior rami of the fifth, sixth, seventh and eighth cervical and first thoracic nerves joining together and forming a network, which stretches from above the clavicle to the axilla.

It gives off five main branches:

(a) *The circumflex nerve.* This passes below the shoulder-blade, then winds round the head of the humerus and supplies the deltoid muscle.

(*b*) *The musculo-cutaneous nerve*. This pierces the coraco-brachialis muscle and runs down the inner side of the arm. It supplies the biceps, coraco-brachialis and brachialis muscles.

(*c*) *The musculo-spiralis nerve*. This winds round the head of the humerus from back to front and supplies the triceps muscle. Below the elbow joint it divides into the radial and posterior interosseous nerves. The radial nerve supplies the skin on the radial side of the forearm and the back of the hand, and the posterior interosseous nerve supplies the extensor and supinator muscles of the wrist and hand.

(*d*) *The ulnar nerve*. This passes down the inner border of the arm and supplies the deep flexor muscles of the wrist, and some small hand muscles.

(*e*) *The median nerve*. This supplies the flexor and pronator muscles of the wrist and fingers.

(3) **The lumbar plexus.** This is formed by the anterior rami of the first four lumbar nerves joining together and forming a network in the psoas muscle. Its branches include the anterior crural or femoral nerve and the obturator nerve.

The femoral nerve passes in the thigh and gives off branches to the anterior and medial extensor thigh muscles. It also gives off the long saphenous nerve which passes down the medial border of the leg to supply the sole of the foot.

The obturator nerve passes out in the thigh through the obturator foramen and supplies the adductor muscles.

(4) **The sacral plexus.** This is formed by the anterior rami of the fourth and fifth lumbar nerves and the upper four sacral nerves. Its branches include the small sciatic nerve to the back of the thigh, the external genitals to the reproductive organs, and the great sciatic nerve, the largest nerve in the body. The great sciatic nerve passes out of the pelvis and down the back of the thigh. About the middle of the thigh it divides into the external and internal popliteal nerves supplying the calf and foot.

Where the groups of nerves forming the brachial plexus and the lumbar plexus are given off, the spinal cord is thickened to form the cervical and lumbar enlargements.

Functions of the spinal cord

The spinal cord is the centre for primitive reflex action (see page 31). If the spinal cord in the lumbar region is injured a person is paralysed in the part of the body below the injury, but reflex action can still take place. For instance, if the foot touches something hot it will be dragged away, although the person is not conscious of any change in the surroundings.

Normally, however, although reflex action takes place involuntarily, the brain is conscious of both the stimulus and the reaction (except in the case of organs supplied by the autonomic system – see page 168).

The sensory neurones which enter the spinal cord may branch and form a synapse with the cell of more than one connector neurone. The axon of a connector neurone may:

(i) Pass through the grey matter to form a synapse with a motor neurone of the same side.

(ii) Pass up or down in the white matter of the spinal cord and then form a synapse with a motor neurone in another segment higher or lower in the spinal cord.

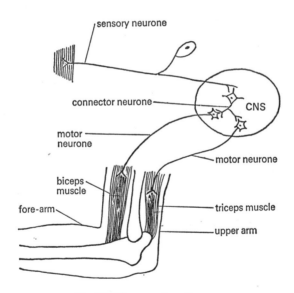

Fig. 92. Diagram of a reflex arc.

(iii) Pass up in the white matter of the spinal cord to form a synapse with other neurones in the brain and so the brain becomes conscious of the stimulation.

Axons taking similar courses are grouped together so that the white matter of the spinal cord is mapped out into definite tracts of axons.

The brain can influence these primitive reflexes of the spinal cord by

160   The nervous system

initiating impulses which pass downwards, so that the white matter of the spinal cord can also be mapped out into tracts of axons descending from the brain.

A connector neurone may form a synapse with more than one motor neurone (see Fig. 92). This is important, because, in each movement, for every muscle which contracts there is a corresponding one which relaxes. For example, flexing the forearm is brought about by the contraction of the biceps muscle. This is accompanied by the relaxation of the triceps muscle, which when it contracts extends the forearm (see Fig. 92). Thus the impulse which passes along the connector neurone and across to the two motor neurones stimulates the biceps and inhibits the triceps. This is known as reciprocal innervation.

The brain

The spinal cord swells out to form the first part of the brain stem called the medulla oblongata. The latter lies just inside the foramen magnum of the occipital bone.

On the posterior side of the medulla is the cerebellum. It has a ridged surface and is divided into two halves, called the right and left hemi-

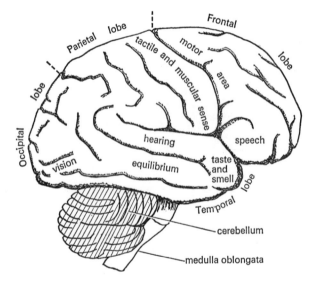

Fig. 93. Diagram of right side of brain showing the sensory and motor area of the cerebral cortex (see page 166).

spheres. If a hemisphere is cut longitudinally it is seen that the grey and white matter are arranged in such a way that they give the impression of tree branches.

The two hemispheres of the cerebellum are joined by a bridge of nerve fibres called the pons or middle peduncle. This runs across in front of the medulla oblongata and so encircles the brain stem.

Two bridges of nerve tissue leave the brain stem below the cerebellum. These join the cerebellum and are known as the inferior peduncles.

The medulla oblongata and the cerebellum together form the hind-brain.

Above the middle peduncle the brain stem swells out on its posterior surface to form four little swellings, the corpora quadrigemina. These, together with the brain stem from which they arise, form the mid-brain. The mid-brain is connected to the cerebellum by the superior peduncles, two more bridges of nerve tissue. In front of the mid-brain the brain stem becomes the thalamus. The brain stem then divides into two to form two peduncles. One passes to each hemisphere to the fore-brain. The two hemispheres form the cerebrum, and the cerebrum and the thalamus together make up the fore-brain.

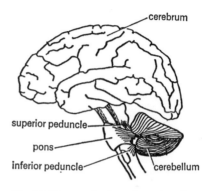

Fig. 94. Diagram to show the connections of the brain.

The cerebral hemispheres fill the top and front portions of the cranial cavity, stretching from above the foramen magnum to the forehead. The two hemispheres are separated by a deep cleft, the longitudinal fissure, at the foot of which runs a bridge of nerve fibres, the corpus callosum, joining the two hemispheres together. The hemispheres are also crossed by other clefts, the largest of which are:

(1) *The fissure of Rolando.* This runs downwards and forwards from about the middle of the longitudinal fissure.

(2) *The fissure of Sylvius.* This runs upwards and backwards from either side of the base of the brain. It forms a deep cleft between the temporal lobe (below the temporal bone) and the rest of the hemisphere.

The grey matter comes to the outside of the hemispheres forming the

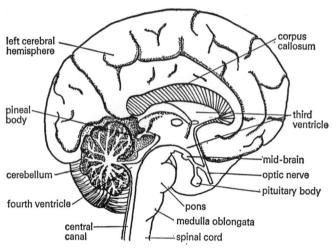

Fig. 95. Longitudinal section of brain.

cerebral cortex. In between the fissures this cortex is thrown into folds or convolutions.

On the floor of the thalamus is a stalk, the infundibulum which is attached to a ductless gland, the pituitary gland (see page 208). In front of the infundibulum is the optic chiasma or crossed optic nerves.

The ventricles of the brain

The small central canal of the spinal cord widens out when it reaches the brain to form a system of cavities called the ventricles.

The fourth ventricle is the name of the cavity inside the medulla. It communicates with the third ventricle, the cavity inside the thalamus, by a narrow canal, the aqueduct of Sylvius, which traverses the mid-brain. This third ventricle communicates with two lateral ventricles, one inside each cerebral hemisphere, by the foramina of Munro.

The ventricles are filled with cerebro-spinal fluid, which is secreted by the arachnoid membrane. It passes from the subarachnoid space into the cavity inside the central nervous system via three holes in the roof of the fourth ventricle.

Cranial nerves

There are 12 pairs of cranial nerves. Unlike the spinal nerves they do not all arise by two roots. A cranial nerve may be made up of sensory neurones, motor neurones, or a mixture of the two.

The sensory nerves. *Of smell:* Nerve I. *The olfactory nerve.* This passes from endings in the mucous membrane within the nose (see Fig. 109) through the holes in the ethmoid bone. It enters the olfactory bulb. Its fibres form synapses with neurones of the olfactory tract which passes to the fore-brain.

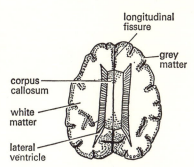

Fig. 96. A section through cerebral hemispheres.

Of sight: Nerve II. *The optic nerve.* This passes from endings in the retina (see Fig. 103) through the optic foramen in the wall of the orbit and enters the cranial cavity. In front of the hypophysis or pituitary gland some of the fibres of the right and left optic nerves go to form the optic chiasma. Inside the brain the fibres then combine with the optic tracts, which pass to the corpora quadrigemina. There they form synapses with more fibres which pass to the visual centre in the occipital lobes of the cerebral hemispheres.

Of hearing and balance: Nerve VIII. *The auditory nerve.* This nerve has two parts, the cochlear nerve, which passes from the organs of Corti (the organs of hearing), and the vestibular nerve, which passes from the semi-circular canals (organs of balance). The two parts unite to form the auditory nerve, which leaves the ear and enters the hind-brain.

The motor nerves. *To the eye:* (*a*) Nerve III. *The oculomotor nerve.* This arises from the mid-brain and supplies the superior, inferior and internal rectus and inferior oblique muscles which move the eyeball; the muscles which raise the eyelid; the constrictor muscles of the pupil; and the ciliary muscles, which alter the shape of the lens.

(*b*) Nerve IV. *The trochlear nerve.* This arises from the mid-brain and supplies the superior oblique muscle.

(*c*) Nerve VI. *The abducens nerve.* This arises from the lower part of the pons and supplies the external rectus muscle of the eye.

To the tongue: Nerve XII. *The hypoglossal nerve.* This arises from the medulla oblongata. It supplies the muscles of the tongue and so controls the movement of the tongue in speech and swallowing.

Mixed nerves. Nerve V. *The trigeminal nerve.* This nerve arises from the front of the pons by two roots. The small ventral root contains motor

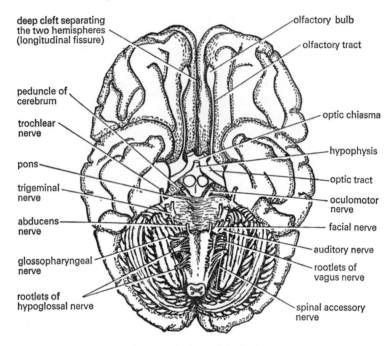

deep cleft separating the two hemispheres (longitudinal fissure)

olfactory bulb

olfactory tract

peduncle of cerebrum

trochlear nerve

pons

trigeminal nerve

abducens nerve

glossopharyngeal nerve

rootlets of hypoglossal nerve

optic chiasma

hypophysis

optic tract

oculomotor nerve

facial nerve

auditory nerve

rootlets of vagus nerve

spinal accessory nerve

Fig. 97. The base of the brain.

fibres, which pass to the muscles of mastication, the sweat and the lachry-mal glands.

The dorsal root has a large ganglion, which is made up of cells of sensory neurones as in a spinal nerve. After the ganglion the nerve divides into three sensory branches:

(1) The ophthalmic nerve, a branch which passes into the orbit and supplies the eyeballs, the upper part of the interior of the nose, and part of the skin.

(2) The maxillary nerve. This crosses the floor of the orbit and then comes out on the face. It supplies the rest of the interior of the nose, the teeth in the upper jaw, and the skin of the face round the nose.

(3) The mandibular nerve. This supplies the front of the tongue, the floor of the mouth, the inside of the cheek, the teeth of the lower jaw, and the skin round the mouth.

Nerve VII. *The facial nerve.* This arises from the lower part of the pons. It is often called the nerve of facial expression, as it supplies motor fibres to the muscles of the face. It contains sensory fibres for the palate and the front part of the tongue, and motor fibres to the salivary glands.

Nerve IX. *The glossopharyngeal nerve.* This arises from the medulla oblongata. It supplies sensory fibres to the hind part of the tongue and pharynx, and motor fibres to the pharyngeal muscles.

Nerve X. *The vagus.* This nerve arises from the medulla oblongata. It descends and passes out of the skull. It then passes down to the abdomen, giving off a number of branches on the way. The most important are:

(1) The laryngeal nerve, supplying the larynx with motor and sensory fibres.

(2) The pharyngeal nerve. This nerve, along with the glossopharyngeal nerve, supplies the pharynx with motor fibres.

(3) The oesophageal nerve, supplying the oesophagus with sensory and motor fibres.

(4) A branch supplying the heart (see page 90).

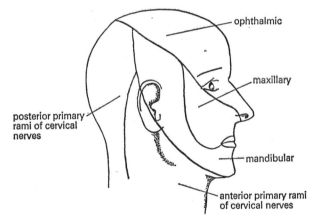

Fig. 98. Diagram to show the areas served by nerves to the skin of the face and scalp.

(5) Sensory branches to the lungs (see page 105), and motor branches to the walls of the bronchi and bronchioles.

(6) Branches to the contents of the abdomen. The vagus supplies most of the gut inside the abdomen and also the digestive glands.

Nerve XI. *The spinal accessory nerve.* This arises from the medulla oblongata and passes out of the skull with the vagus; it sends some fibres to join the vagus and then supplies the sternomastoid and trapezius muscles with motor fibres.

Functions of the brain

The fore-brain

The cerebral cortex contains the higher centres for consciousness, reasoning, memory and thought, and it also controls all voluntary movement (including speech) and the co-ordination of the muscles of the whole body.

It is possible to divide the cortex up into a sensory area, which receives and deals with all sensory impulses, and a motor area, which initiates impulses which bring about voluntary movement.

Both the sensory area and the motor area can be mapped out as shown in Fig. 93.

Once a soldier was wounded in such a way that his skull was torn and his brain partially exposed. It was found that when an electric current was applied to a particular spot on what is known as the motor area a certain movement was produced in a particular muscular area, and it was different from the movement produced when any other spot was stimulated. At the same time it was noticed that the stimulation of the motor area on one side brought about a movement of the muscles on the other side of the body, and vice versa.

Many sensory impulses reach the sensory cortex from the lower part of the body. Sensory neurones pass into the spinal cord and carry impulses from the skin, muscles, tendons, joints, etc. They form a synapse with connector neurones, whose axons cross to the other side of the central nervous system, and then pass to the thalamus, which is the seat of a vague consciousness capable of appreciating pain, rough contact and extremes of temperature. A dog with cerebral hemispheres removed will still carry out conscious actions, but in a clumsy kind of way. If it sees food at the other side of the room it will go clumsily towards it and eat it. A dog with the entire fore-brain removed will only carry out reflex actions. It will eat food, but only if the food be placed in its mouth.

In the thalamus the connector neurones form a synapse with other neurones whose axons pass to the sensory area of the cerebral cortex. Here there is a finer appreciation of the stimuli including light touch, slight change in temperature and muscle sense.

As a result of these tracts conveying impulses, the brain becomes conscious of any changes in the environment or position of the parts of the body.

Besides the tracts of sensory fibres passing to the cerebral cortex, tracts of motor fibres leave the motor cortex. Some pass down the brain stem to the medulla oblongata, where they form two pyramidal-shaped tracts. In the medulla they cross and then pass on to the anterior horns of grey matter in the spinal cord. There they form synapses with motor neurones, and as a result the cerebral cortex controls muscular movement.

Since the descending fibres cross, the left cerebral cortex controls the muscles on the right side of the body and vice versa. Thus if the left cerebral cortex is injured the right side of the body is paralysed.

Other tracts of descending fibres pass from the cerebral cortex to the cerebellum.

The mid-brain

This is the centre for reflexes connected with the eye. For example, when something comes near the eye an impulse passes along the optic nerve to the mid-brain, then across connector neurones and down the motor nerve, which supplies the muscles of the eyelid, and the eye blinks.

The fibres of the optic nerve also form synapses in the mid-brain with connector neurones which pass to the sensory area of the cerebral cortex, and the brain becomes conscious of the appearance of the surroundings.

The hind-brain

The medulla oblongata contains the centres for reflexes which control respiratory movements (the respiratory centre), the rate of the heart (cardiac centre), swallowing, vomiting, etc.

The cerebellum is the centre for reflexes, which bring about muscular co-ordination and maintenance of balance.

Sensory impulses telling the positions of the muscles, tendons and joints not only pass along connector fibres to the thalamus, but also pass along different tracts to the cerebellum.

The vestibular fibres from the vestibular apparatus of the ear or organ of balance form synapses in the medulla oblongata with neurones which pass to the cerebellum.

The cerebellum is connected up with the cerebral hemispheres by tracts of fibres which convey impulses from the cerebellum to the cerebral cortex, and by tracts which convey impulses from the cerebral cortex to the cerebellum.

The fibres between the cerebellum and the cerebral cortex cross, and therefore one side of the cerebellum is connected with the opposite cerebral hemisphere. By means of these connecting fibres the cerebral hemispheres can influence the working of the cerebellum, which regulates the balance and equilibrium of the body. For example, when a child first begins to ride a bicycle it thinks about every movement, impulses being initiated in the motor cortex. Eventually with practice the muscles begin to act reflexly and the process becomes automatic, impulses no longer being initiated in the cerebral cortex.

Conditioned reflex

It was pointed out in Chapter 12 that there may be a psychic flow of saliva at the sight or smell of food. This is an example of a conditioned (acquired) reflex. It only takes place if the person is familiar with the food and has come to associate the sight or smell of the food with a particular kind of taste, when pathways become established between his cortical centres for sight and smell, and his salivary glands.

In everyday life many simple reflexes become conditioned in this way and form the basis of habit formation, learning and training.

The autonomic nervous system

This part of the nervous system supplies the organs of the body which are not under the control of the will.

The autonomic system is divided into the sympathetic system and the parasympathetic system.

Sympathetic system (STIMULATED BY FEAR)

From the thoracic and upper three lumbar spinal nerves, after the union of the dorsal and ventral roots, branches or white rami communicantes are given off one from each nerve.

Each ramus passes to a sympathetic nerve ganglion. There are two rows of such ganglia joined by a nerve chain lying one on either side of the vertebral column. A sensory neurone of the sympathetic system passes along a spinal nerve or white ramus and enters the spinal cord by the posterior root, its cell body lying in the posterior ganglion. In the grey matter of the spinal cord the sensory neurone forms a synapse with a connector neurone which passes out from the cord along the anterior root and white ramus communicans. Sympathetic connector neurones may then do one of three things. They may form synapses with sympathetic motor neurones in the sympathetic ganglia at the same level of the body; they may run up and down the body to form synapses in the sympathetic ganglia of a different level, in this way helping to form the sympathetic nerve chain or they may run out from the sympathetic chain to form synapses in outlying ganglia. The motor neurones or post-ganglionic neurones are non-medullated. Those from the ganglia of the sympathetic chain form the grey rami communicantes which join the spinal nerves and pass to the organs, chiefly blood vessels and sweat glands, in the spinal nerves.

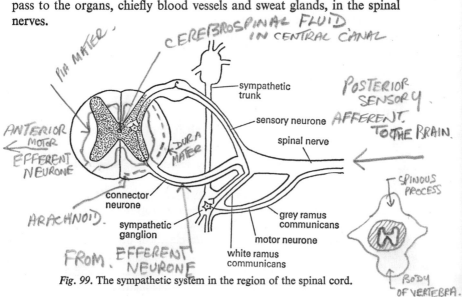

Fig. 99. The sympathetic system in the region of the spinal cord.

Others from the outlying ganglia form sympathetic nerve plexuses or nerve nets. From these nerves pass out to supply the viscera.

The chief plexuses are:

(i) *The cardiac plexus.* This lies near the base of the heart and sends branches to the heart and lungs.

(ii) *The solar plexus.* This lies behind the stomach and supplies the abdominal viscera.

(iii) *The hypogastric plexus.* This lies in front of the sacrum and supplies the pelvic organs.

Parasympathetic system

This is made up of connector fibres, which pass in some of the cranial nerves and the second and third sacral nerves. These fibres relay with motor fibres within the organ supplied. The autonomic fibres in the sacral region combine to form the pelvic nerves.

In most cases an organ receives fibres from both the sympathetic and parasympathetic system, and the two systems generally have an antagonistic action, e.g.

(*a*) The heart receives accelerator fibres from the sympathetic system, the fibres passing from a ganglion in the sympathetic chain of the neck region, and inhibiting fibres from the parasympathetic system in the vagus nerve (see page 90).

(b) The muscles of the bladder receive motor fibres from the para- sympathetic system in the pelvic nerves and inhibiting fibres along the sympathetic system, the fibres passing from the hypogastric plexus (see page 149).

PRACTICAL

1 Dissect out the spinal nerves of the rabbit from the ventral side.

2 Examine the cut end of part of the spinal cord of an ox.

3 Examine stained and mounted transverse sections of different regions of the spinal cord.

4 Reflex action. Demonstrate the knee-jerk reflex. A person sits with one knee over the other. The ligament below the knee-cap is struck lightly with the side of the hand and the leg jerks upwards. The strength of this response, which is more marked if the fists are clenched, depends on the nervous condition of the body.

5 Examine the brain of a sheep in detail. Cut it longitudinally and transversely, and note the distribution of the grey and white matter.

17 Organs of special sense

In the higher animals, including man, parts of the nervous system are adapted to receive and interpret special kinds of stimuli. These are the special sense organs and include:

(1) The eye, which perceives light rays.

(2) The ear, which perceives sound waves and acts as an organ of balance, being sensitive to the position of the head.

(3) The organs of taste in the mouth.

(4) The sensory epithelium of the nose, which is sensitive to smells.

(5) The tactile corpuscles situated in the skin and muscles. These are sensitive to touch.

The eye

The eyeball lies in a socket known as the orbit. The orbit is lined with a cushion of fat which protects the eye from mechanical shocks. In front the eye is protected by the eyelids, which are edged with small hairs known as eyelashes.

Lining the eyelids is a mucous membrane. It is the continuation of the delicate membrane which covers the exposed part of the eye. Over the cornea it consists only of epithelium. At the margins of the eyelids it is continuous with the skin.

The lachrymal gland lies on the outer and upper corner of the orbit (see Fig. 100). It is continually secreting a fluid which passes over the outside of the eye, keeping it moist and removing any dust particles. The fluid drains away into a duct, the naso-lachrymal duct, which passes from the inner side of the eye to the nose. When this duct is blocked the eye waters. When the lachrymal gland is stimulated, its secretion increases and tears are formed.

On the outside of the eyeball is the sclerotic, a protective coat of strong connective tissue. The sclerotic forms the white of the eye and bulges in front of the eye to form the transparent cornea, which protects the lens (see Fig. 102).

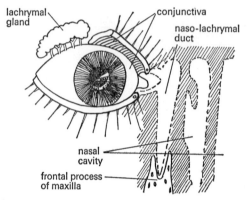

Fig. 100. Diagram of the eye to show lachrymal gland.

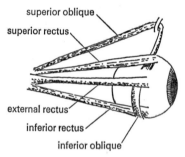

Fig. 101. Muscles of the orbit.

On the outside of the sclerotic and inside the orbit six muscles are attached which move the eyeball. The six muscles are:

The superior rectus, which moves the eye upwards.

The inferior rectus, which moves the eye downwards.

The internal rectus, which moves the eye inwards.

The external rectus, which moves the eye outwards.

The superior oblique, which moves the eye obliquely downwards and outwards.

The inferior oblique, which moves the eye obliquely upwards and outwards.

Inside the sclerotic is a dark-coloured coat called the choroid. It contains a network of blood capillaries, with black pigment cells in the meshes. The pigment cells absorb light, so that the choroid forms a light-absorbing layer. In many animals such as cats there is iridescent pigment in this layer, and the eyes shine in the dark.

In front of the eye, where the sclerotic becomes the transparent cornea the choroid leaves the sclerotic and forms the iris, the part of the eye which varies in colour from person to person. In the centre of the iris is a small hole known as the pupil.

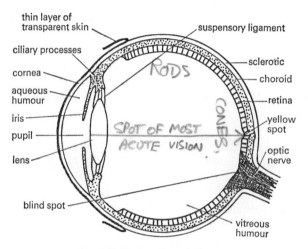

Fig. 102. Section through the eye.

Within the choroid is the light-sensitive coat of the eye, the retina. The retina consists of several layers. Next the choroid is a layer of pigment cells; next to these cells is a layer consisting of two types of cells called the rods and the cones (see Fig. 103). The rods contain a purple-red colouring matter called visual purple, which is bleached on exposure to light.

The nuclei of the rods and cones lie in projections from the cells. From these projections, axons pass inwards and form synapses with dendrites of a row of oval cells. The axons of the latter pass inwards and many form synapses with the dendrites of other nerve cells, the ganglion cells. The axons of these leave the surface of the retina and become the optic nerve.

Where the optic nerve pierces the retina there are no rods and cones, and a blind spot insensitive to light is present.

In the centre of the retina is a yellowish spot. Here the retina is very

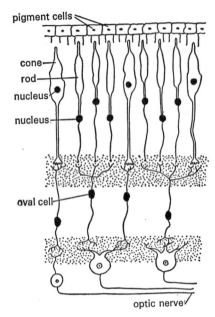

pigment cells

cone

rod

nucleus

nucleus

oval cell

optic nerve

Fig. 103. Section through the retina.

thin, consisting chiefly of cones; in the centre of the spot there are no rods. Passing from the centre to the periphery the cones become progressively less numerous and the rods more so.

Lying behind the iris suspended in position by an annular ligament, the suspensory ligament, is the biconvex lens. The suspensory ligament blends with the capsule covering the lens and extends outwards to be attached to ciliary processes formed by the folds of the choroid and through them to the ciliary muscle. The latter forms a circular band on the outer surface of the front part of the choroid. When it contracts it draws forward the ciliary processes and relaxes the ligaments. The lens which is elastic becomes more convex.

Between the lens and the cornea there is a cavity filled with a fluid, the aqueous humour; the cavity within the retina is filled with vitreous humour, a jelly-like substance. These 'humours' give the eye its shape and keep the organ firm.

Functions of the eye

The eye is the organ of the body especially adapted for receiving light stimuli and converting them into nervous impulses, which it transmits to the brain via the optic nerve.

The lens, cornea and humours of the eye function in the same way as the lens of a camera. In a normal eye the light rays come to a focus on the retina in such a way that an inverted image of the object is received. Impulses pass to the brain, which interprets the image, and the objects are perceived the correct way up.

The amount of light falling on the lens depends on the size of the pupil. This can be regulated by the iris, which possesses two sets of muscles, the radial and the circular.

When the radial muscles contract the pupil enlarges; when the circular muscles contract the pupil becomes smaller. When the latter happens the amount of light entering the eyeball is reduced, but the image is seen more distinctly.

The pupil is small when the light is bright or when the eye is examining a small object near at hand. It is large when the light is dim or the eye is focused on distant objects. It corresponds to the 'stop' of a camera.

Rods and cones. Since rods and cones are not present on the blind spot, which is insensitive to light, and since the yellow spot consisting only of rods and cones is very sensitive, it follows that the rods and cones form the part of the retina which receive the light stimuli. When the eye is examining a fine object the image is focused on the yellow spot. As the yellow spot consists chiefly of cones, it is probable that the cones are for accurate vision.

The rods contain a dye called visual purple, which is bleached by light. They are sensitive to light of lower intensity than the cones and it is probable that they are adapted for twilight vision. They are not sensitive to different wave-lengths and therefore do not give rise to any colour sensation but only black, white and grey.

Colour vision

Several theories of colour vision have been put forward. According to Young-Helmholtz's Theory three kinds of cones exist in the retina, each type containing a substance sensitive to light of a different colour, one type being stimulated by red light, another by green light and the third by blue light, while white light stimulates the three kinds of cones. Likewise, yellow light, for example, stimulates an equal number of red and green cones.

Colour blindness

Man is affected by different types of colour blindness:

Total colour blindness. The sufferer is unable to perceive colours and sees objects in various tones of grey. In the retina there are no cones.

Partial colour blindness. The colour blindness is limited to certain colours, usually red and green. On Young-Helmholtz's Theory of colour vision partial colour blindness is due to the absence of one of three types of cones normally present in the retina of the eye.

After-image

Positive. When a bright object is looked at and the eyes closed the image persists. This is known as the after-image, and this phenomenon allows a continuous image to be seen, when in fact there are a series of different images, e.g. in moving pictures the image appears one of continuous motion, while actually there are as many as twenty-four pictures shown on the screen in a second. It is probably due to changes caused by the light on the retina outlasting the stimulus.

Negative. If, on the other hand, after staring at a bright light one then turns one's eyes to a light surface, a dark image is seen. If the first image is coloured, the after-image appears to be in the complementary colour.

Accommodation

A lens bends rays of light towards its thickest part; the greater the curvature of the surface, the greater the bending. For a lens of fixed curvature, rays from a near object are brought to a focus farther from the lens than those from a distant object.

In an eye, the distance of the image on the retina from the lens is fixed, but the lens in the eye can alter its curvature; this is called 'accommodation'. It has already been shown how the ciliary muscle alters the shape of the lens. Under resting conditions these muscles are relaxed, the suspensory ligaments are taut and the lens is flat and adapted for long-distance vision. When the ciliary muscles contract, the suspensory ligaments become less taut and the lens becomes thicker and more convex, and the eye is adapted for near vision.

Long and short sight

Long sight. In some eyes, the image of either a near or a distant object is not formed on the retina. If the eye cannot see near objects clearly, it is

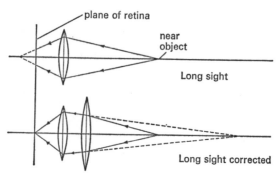

Fig. 104(a). Correction of long sight.

said to suffer from 'long-sight' (hypermetropia). This is corrected by using a spectacle lens, the purpose of which is to form an image (of the near object) in such a position that the eye can see it clearly, i.e. farther from the eye than the object.

The image of the near object is now sufficiently distant for the eye to 'accommodate' it.

Short sight (myopia) is the inability to 'accommodate' distant objects.

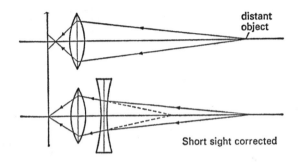

Fig. 104(b). Correction of short sight.

The image of the distant object is now sufficiently near for the eye to 'accommodate' it.

Presbyopia is the loss of accommodation because of the hardening of the ciliary muscles, or loss of elasticity of the lens, in middle and old age. There is only a limited range of distances of the object for which the image can be focused on the retina; spectacle lenses, either two pairs or bifocals are needed.

Binocular vision and squinting

When normal eyes are directed at an object, images only approximately the same are produced on each retina. Any retinal defects of one eye are masked, and the blind spots are not detected.

The position of the eyes is adjusted to allow the image to fall on corresponding parts of both retinae. This adjustment is brought about by means of the external eye muscles, the nearer the object the greater the convergence of the two eyes. Normally this adjustment is well balanced, but in the condition of squint two images are produced.

Since approximately similar images are focused on the two retinae, one retina sees a little more to the right, the other a little more to the left. This enables the brain to perceive in three dimensions and the objects appear solid.

The ear

The ear is the organ of the body especially adapted for perceiving sound waves. It is also the body's balancing organ.

The part of the ear on the outside of the head is the pinna. This consists of elastic cartilage covered with skin. There are muscles attached

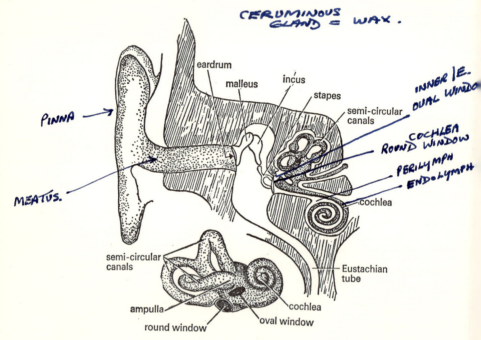

Fig. 105. Section through the ear and bony internal ear removed from the temporal bone.

to the pinna, but they usually have no function. Certain animals, e.g. rabbit, however, can move their pinnae towards the direction of the sound, and so collect the sound waves.

The sound waves are conveyed inward through a curved canal called the external auditory meatus. It is about an inch long. Its walls secrete wax and its external opening is protected by hairs. The pinna and external auditory meatus together form the external ear.

At the base of the meatus is the tympanic membrane or ear-drum which separates the external from the middle ear.

The middle ear is a cavity lined by mucous membrane in the temporal bone. From its floor the Eustachian tube passes downwards and leads into the pharynx by an opening which is closed except when swallowing is taking place. Then the air in the Eustachian tube is renewed and the air pressure in the middle ear is kept at atmospheric pressure. During a cold the wall of the Eustachian tube may become inflamed and the tube blocked, then deafness may be produced, since the pressure can no longer be regulated.

On the inner wall of the middle ear in the bone there are two openings, the fenestra ovalis or oval window, and the fenestra rotunda or round window. Both these openings are covered with membrane, and separate the middle from the inner ear.

Stretching from the tympanic membrane to the fenestra ovalis is a string of three bones – the malleus, incus and stapes.

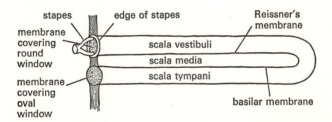

Fig. 106. Vertical section through the cochlea, straightened out.

The malleus is hammer shaped, the handle of the hammer being attached to the tympanic membrane. The head of the hammer articulates with the incus or anvil bone. The other end of the incus articulates with the small end of the stapes or stirrup bone, while the larger end of the stapes is attached to the membrane stretched across the oval window. This chain of bones in the middle ear is kept in position by ligaments.

The inner ear consists of the bony labyrinth, a system of cavities in the petrous portion of the temporal bone.

It has a central part, the vestibule, from which three semi-circular canals in the bone and a spiral tube called the cochlea, are given off.

The three semi-circular canals project from the back of the vestibule. They are so arranged in three planes that they are at right angles to one another.

The cochlea passes from the front of the vestibule and coils round a central bony pillar, the modiolus. Up the centre of the latter, which is hollow, passes the cochlear part of the auditory nerve.

The cochlea is divided incompletely into two by a ledge of bone which winds round the modiolus. Stretching from the tip of this ledge are two membranes, the basilar membrane and Reissner's membrane, which extend to the outer wall of the tube, the basilar membrane meeting the canal wall a short distance below the Reissner's. In this way the membranous labyrinth is formed and the spiral canal divided into three spiral compartments:

(1) Scala vestibuli, above the Reissner's membrane.

(2) Scala tympani, below the basilar membrane.

(3) Scala media, between the basilar and Reissner's membrane.

The scala media is filled with endolymph and the scala vestibuli and scala tympani with perilymph.

The scala tympani begins at the round window and the scala vestibuli at the oval window, both windows being closed by a membrane separating the middle from the inner ear.

Within the vestibule there are two membranous sacs, the utricle and saccule.

The saccule gives off the membranous cochlea, the utricle the membranous semi-circular canals. At one end of each semi-circular canal there is a swelling or ampulla.

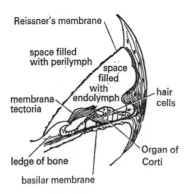

Fig. 107. Section through one spiral of cochlea.

The sensory epithelium in the cochlea. The basilar membrane supports a number of columnar epithelial cells, the rods of Corti, and on either side of these some hair cells. The neurones of the auditory nerve end against these hair cells. The rods and hair cells together form the organ of Corti, which is roofed over by a membrane called the membrana tectoria.

Hearing

When sound waves pass down the external auditory meatus they set up vibrations in the tympanic membrane. The vibrations are then transmitted by the chain of bones to the membrane covering the fenestra ovalis. The vibrating of this membrane sets up waves in the perilymph surrounding the membranous canal of the cochlea. The perilymph then beats on the membrane and waves are set up in the endolymph. These movements are transmitted to the basilar membrane whose movement stimulates the hair cells and impulses are conveyed to the brain along the auditory nerve.

Balance

Inside the ampulla of the semi-circular canals is a sensory epithelium, consisting of columnar epithelial cells. Some of these, the sensory cells, have stiff hairs attached on the outside, and have fibres of the eighth nerve passing into them on the inside; the others are the supporting cells. The stiff hairs of the sensory cells project into the endolymph which fills the canals.

When the head moves the endolymph within the membranous semi-circular canals moves at a different speed and so stimulates the hair cells. Impulses are then initiated which pass to the brain.

Since the semi-circular canals lie in three planes, some hair cells are stimulated in whatever direction the head is moved, so that by means of these sense organs a person can determine the direction and approximate speed of any movement, even though he be blindfolded without his feet touching the ground.

If the semi-circular canals are diseased the patient suffers great disturbances of equilibrium, having a reeling gait. He may also have a feeling of giddiness or sickness. If a man with diseased semi-circular canals is blindfolded and then dives into water, he will not be able to tell whether he is moving up or down in the water.

Within the saccule and utricle there are similar sensory epithelia, but in this case among the stiff hairs of the hair cells are calcareous bodies embedded in the jelly. These change their position when the position of the head is moved with respect to gravity, and so stimulate the hair cells. As a result impulses are initiated which help to determine the position of the head.

It follows from the above that the ear is not only an organ for perceiving and transmitting to the brain sound waves from the outside world, but it also transmits to the brain impulses from within the body regarding the movement and position of the head.

Taste

The sense of taste resides in the mucous membrane of the tongue and palate, and to a lesser degree the pharynx. The tongue is a muscular organ which assists in the mastication and swallowing of food.

The tongue muscles lie in two groups:

(1) The *intrinsic* muscle group. These muscles bring about the finer tongue movements.

(2) The *extrinsic* muscle group. These muscles bring about the larger movements such as swallowing, and attach the tongue to the back and floor of the mouth.

The taste organs or taste buds are situated on the walls of projections or papillae on the mucous membrane of the tongue; they are also found in the mucous membrane of the palate and pharynx.

A taste bud consists of a collection of cells formed into kind a of bud. In the centre of the bud are cells with long hairs, and amongst these, fibres of the taste nerve branch.

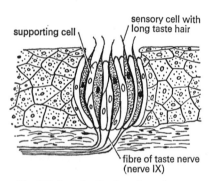

sensory cell with long taste hair

supporting cell

fibre of taste nerve (nerve IX)

Fig. 108. Longitudinal section of a taste bud (×400).

The tongue papillae (see Fig. 71) are of three kinds:

(1) *The filiform papillae.* These are found all over the surface of the tongue.

(2) *The fungiform papillae.* These are mushroom-shaped and are found on the top and side of the tongue.

(3) *The circumvallate papillae.* These are longer than the others and are surrounded by a kind of valley. There are only about ten of them altogether, placed at the base of the tongue.

The majority of the taste buds are found on the fungiform and circumvallate papillae, the filiform papillae being concerned more with the sense of touch.

The taste buds perceive the nature of the substances dissolved in saliva, substances only being tasted if and when they go into solution, their molecules impinging on the hairs of the taste-bud cells.

There are only four kinds of taste: sweet, sour, bitter and salt, other 'tastes' being really smells. For this reason a cold in the nose often makes food appear tasteless. To a man with his nose blocked an onion would taste no different from an artichoke.

The whole surface of the tongue seems to be equally sensitive to saltness, but not to the other tastes. The tip of the tongue seems to be sensitive to sweetness; the back to bitterness, and the sides to sourness.

Smell

The sense of smell is situated in the inside of the nose. It is relatively feeble in man compared with the majority of animals, especially the dog.

The nose is divided into two nasal cavities, from the outer side of which three conchae project.

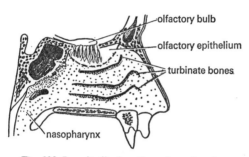

Fig. 109. Longitudinal section of nasal cavity.

The conchae and the rest of the inside of the nasal cavity are covered with a mucous membrane. The mucous membrane, which surrounds the path in the lower part of the nasal cavity taken by the air, has cilia which remove the dirt and germs from the air breathed in (see page 98). The mucous membrane in the upper part or olfactory part has an epithelium consisting of several layers. Some of these cells are stimulated by molecules of substances taken in with the air. An impulse then passes up the olfactory nerve, whose fibres run into these epithelial cells, and a smell is perceived. To reach the sensory epithelium the olfactory nerve passes through small holes in the ethmoid bone. The olfactory receptors adapt very quickly, that is, they soon cease to respond to some particular stimulus and a person may become oblivious of a strong smell.

If the molecules of a substance do not stimulate the epithelial cells, the substance has no smell. Sometimes when one has a cold in the head the sense of smell may disappear and normally strong-smelling substances cannot be sensed.

Touch

By means of sense organs in the skin the body is conscious of changes in the surroundings. By them the body may feel light touches, pain, temperature changes and pressure.

Light touch. Light touches are felt by the Meissner's corpuscles, which are oval bodies consisting of connective tissue around the end of a nerve fibre.

Pain. Pain is felt by nerve fibres which end amongst the cells in the deeper layer of the skin. These fibres, which feel only pain, lie very close together. They are closer together in some parts of the skin than in others, so that some parts are more sensitive to pain.

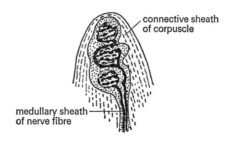

Fig. 110. Meissner's corpuscle (× 200).

These pain nerve endings are also found in membranes within the body, but not in solid organs. As a result some organs, e.g. the liver, can be cut without causing pain.

If an internal organ is disordered in any way a pain may be felt in the membrane surrounding it, or in any other part of the body supplied by the same nerve. For example, a pain in the left arm may indicate some disease of the heart.

Heat and cold. The organs which perceive heat and cold have not been discovered, though they must exist. If the skin is touched with a hot and with a cold needle some spots are found sensitive to heat, some to cold.

Pressure. Pressure on the skin is perceived by Pacinian corpuscles, which lie in the subcutaneous fat, especially in the hand and foot. A corpuscle consists of a nerve fibre surrounded by connective tissue arranged in concentric layers. Between the connective tissue and fibre is a thin layer of protoplasm.

By means of sense organs in the joints, muscles and tendons, the body is conscious of the position of the limbs and the state of contraction of the muscles.

Position of limbs. Pacinian corpuscles are found in the joint membranes. By means of pressure on these, impulses are set up which give information about limb position and so help balance.

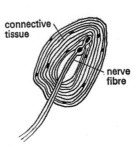

Fig. 111. Pacinian corpuscle (×100).

Muscle sense. In the muscles and tendons are nerve endings which are stimulated by a contracting muscle. Impulses set up give information about the degree of muscular contraction.

PRACTICAL

Eye

1 Examine the eye of the rabbit *in situ*. Note the eye muscles.

2 Examine a model eye.

3 Dissect a bullock's eye. Remove the lens and compare it with a glass biconvex lens.

4 Take another bullock's eye. Cut out a window in the back of the eye. Look through this window at a candle on the other side of the eye. An inverted image of the candle is seen. Place a piece of tissue paper over the window and on to it focus the image of the candle.

5 Examine a prepared slide of the retina of the eye.

6 Movements of the iris and convergence of the eye.
  (*a*) Notice the size of the pupils of the eyes of a person sitting in a good light. Cover over one of his eyes for 3 or 4 seconds and then remove the cover quickly. Notice that the pupils contract rapidly.
  (*b*) Notice the convergence of the eyes and the contraction of the pupils when a person looks from a distant to a near object.

7 Accommodation.
  (*a*) Hold up two pencils, one near the eye and one farther away. Stare at the near pencil and notice that the far one appears blurred. Stare at the far pencil and the near one appears blurred.
  (*b*) Open the eyes wide and stare at print in a book. It appears blurred. When the eyes are wide the ciliary muscles are relaxed and the eye focused for distant vision.

8 Demonstrate the blind spot. Make a black cross and a dot on a piece of paper,

G

the dot about three inches to the right of the cross. Hold the paper about 18 in away from the eyes. Close the right eye and stare at the spot with the left eye. Bring the paper slowly towards the eyes. At one point the cross will disappear, but will soon reappear. The cross disappears when its image falls on the blind spot of the left retina.

9 Show that the two eyes do not see identical images. Place a black mark on the right side of a white cylinder of paper. Close the right eye and move the head until the black spot can just be seen by the left eye. Open the right eye and notice that some of the paper beyond the mark can now be seen.

10 Eye tiredness. Look at a red square fixedly for some time and then at a red object. The eye becomes red tired and the object appears black. Look at a red square and then at a white wall, a greenish-blue square is seen. The eye becomes red tired and the colour complementary to red is seen. This phenomenon is known as the negative after-image.

11 Positive after-image. In a dark room look at a candle flame and then blow it out. The image of the candle persists for a fraction of a second.

Ear

Examine a model of an ear.

Taste

Place a drop of (1) salt solution, (2) sugar solution, (3) quinine solution, (4) lemon-juice, on the tip and then on the sides and on the back of the tongue. Note the sensations.

Touch, etc.

1 Take a pair of dividers and experiment on a blindfolded person. Put the two points of the dividers $\frac{1}{8}$–$\frac{1}{4}$ in apart and prick the person lightly on the palm of the hand sometimes with one point, sometimes with two. The person should state whether he feels one or two pricks. Repeat on different parts of the skin. In this way determine the relative sensitivity to touch of the skin on different parts of the body.

2 Take a very fine metal knitting-needle with a blunt point and place the end in hot water. Touch different spots on the skin of the arm with this needle and mark the spots at which heat is felt.

Place the end of the needle in very cold water and repeat, marking the spots at which cold is felt. Touch different spots on the skin of the arm with a fine bristle until the bristle bends. Mark the touch spots.

Repeat the experiment with a sharp needle. At some points, but not all, pain will be felt. Mark the pain spots.

18 Reproduction

When an Amoeba reaches the adult size it divides to form two 'daughter' Amoebae. These in their turn divide. This formation of new organisms similar to the parent or parents is known as reproduction, and is characteristic of all living organisms.

In the higher animals, composed of masses of cells, there is a division of labour, and certain cells, the germ cells, are set aside for the continuance of the race. These germ cells are protected by the body or somatic cells.

Reproduction in the higher animals is always sexual. A fertilized egg is produced by the fusion of two cells. One is called the female germ cell or ovum, and the other the male germ cell or spermatozoon. The fertilized egg or zygote, when formed, is one-celled, but it begins to divide up to form a new individual consisting of masses of cells. The majority of these cells lose their power of division and become highly specialized to enable them to carry out their own particular function. Some retain their primitive characteristics and their power of division. These are the germ cells which, unlike the somatic cells, are potentially immortal.

When the cells in the body divide the nucleus goes through a series of complicated stages known as mitosis.

The 'resting' nucleus contains chromatin, a material which rapidly takes up certain stains, aggregated into a number of short rods called chromosomes which begin to show up when the cell is about to divide. The number of chromosomes for any given species is constant (the number is represented by $2n$, where n varies from species to species).

Shortly after the chromosomes appear they become shorter, and the centrosome, a small body found outside the nucleus in most animal cells, divides into two. The new centrosomes travel to opposite ends of the cell and between them the protoplasm becomes modified to form a spindle of fine fibres. The nuclear membrane disintegrates and the chromosomes, which are split lengthwise, pass into the cavity of the spindle. They then arrange themselves on the threads at the equator of the spindle, and one half of each chromosome known as a chromatid

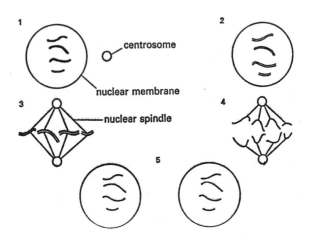

Fig. 112. Diagram of somatic cell division showing four of the 2*n* chromosomes.

1. Four chromosomes have appeared in the cell nucleus, and the centrosome outside it.
2. The chromosomes have split into two along their lengths.
3. The membrane has disappeared and the chromosomes have arranged themselves at the equator of the spindle which formed when the centrosome split into two.
4. The chromosomes are travelling towards the two centrosomes.
5. Two new nuclei have been formed and each has become surrounded by a nuclear membrane.

passes towards each centrosome and becomes a new chromosome. Near the centrosome the chromosomes become obscured, and each of the two groups forms a nucleus surrounded by a nuclear membrane. The cytoplasm then divides into two, and two cells are formed in the place of the original one. It should be noticed that the chromosome number is not altered by this division, the somatic cells still containing 2*n* chromosomes.

Reduction division or meiosis

When in sexual reproduction there is a fusion of two cells, some adjustment has to take place, otherwise the fertilized egg would contain $4n$ chromosomes.

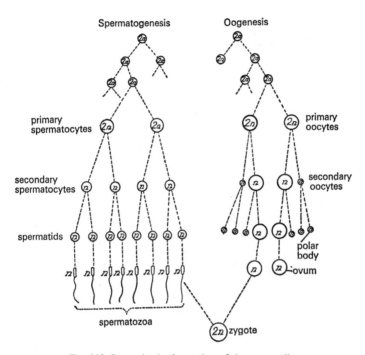

Fig. 113. Stages in the formation of the germ cells.

In one stage of the development of germ cells a special type of nuclear division known as reduction division takes place. This results in the mature ovum and the mature sperm containing n chromosomes each. When fusion takes place the fertilized egg or zygote contains $2n$ chromosomes or the somatic number.

Spermatogenesis, the formation of sperm cells, and oogenesis, the formation of ova, take place in the following stages:

In the male the mother cells of the spermatozoa, the primary spermatocytes, which contain $2n$ chromosomes, divide twice, producing four cells, the spermatids. The first division is a reduction division. In this process the chromosomes arrange themselves at the equator of the spindle in pairs. Although they are split the two halves or chromatids do not separate but the members of each chromosome pair separate, one going towards one centrosome, the other towards the other centrosome. Therefore the new nuclei have one member of each pair of chromosomes, only n chromosomes altogether. In the second division the chromatids in both nuclei divide; four spermatids form, each consisting of n chromosomes, nuclear membrane and cytoplasm. The spermatids develop into the spermatozoa, which have small heads and long tails.

In the female similar changes take place. The primary oocytes or mother cells of the ova containing $2n$ chromosomes divide into two by a reduction division, to form the secondary oocytes. One of these cells, known as the first polar body, is very small. The two cells again divide, this time by ordinary division. The large cell forms a large egg cell or ovum, and a small cell, the second polar body. The first polar body divides into two. Eventually the three polar bodies formed disintegrate.

The determination of sex

In the nucleus of a body cell, the chromosomes are in pairs. One of these pairs consists of the sex chromosomes. In the majority of animals, including man, the two sex chromosomes of the female are identical and called the X chromosomes; the sex chromosomes of the male are not identical; there is a single X chromosome, and a much smaller chromosome called the Y chromosome. Consequently in the formation of the germ cells half the spermatozoa contain a Y chromosome; the other half and all the ova contain an X chromosome. On fertilization, when an X spermatozoon fuses with an ovum a female individual is produced, but when a Y spermatozoon fuses with an ovum a male individual is produced.

Hereditary characteristics

It is now accepted that the chromosomes are the bearers of hereditary characteristics, in other words the characters which an offspring inherits from its parents. For a full description, see Chapter 20 on Inheritance.

Reproduction in man

In all the higher animals the reproductive organs may be divided into the essential organs which produce the germ cells, the spermatozoa in the male, and the ova in the female, and the accessory organs which enable the sperm to reach the ova for fertilization and also protect the developing egg.

Essential organs of the male

The two testes are the essential organs of the male. They develop in the abdominal cavity, but during foetal life they descend through the inguinal canals or Poupart's canals (see page 54) into the scrota, small sacs situated at the outside lower end of the abdomen. Each testis as it descends

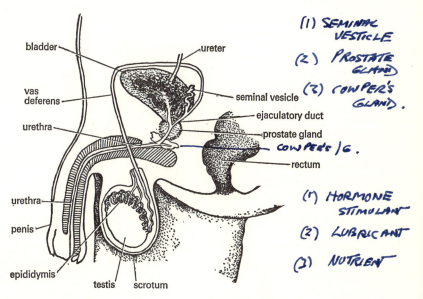

Fig. 114. Section through the male pelvis.

carries with it blood vessels, nerves and its duct or vas deferens, which together form the spermatic cord.

After birth the spermatic cord lies in the inguinal canal and suspends the testis in the scrotum, the left testis hanging down a little lower than the right. The testes enlarge at puberty and begin producing sperms and semen.

A testis is made up of a collection of unbranched convoluted tubules called the seminal tubules, lined with cubical epithelium. The tubules are supported by a framework of fibrous connective tissue, which divides the testis up into compartments. Between these are groups of interstitial cells.

Several tubules unite to form a straight tubule, which in turn unites with many others to form a network. This ends in small ducts, the vasa efferentia, which join up to form a coiled tube, the epididymis, lying

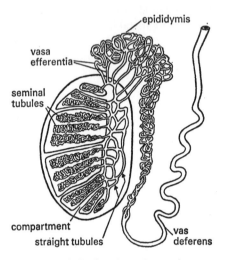

Fig. 115. Section through a testis.

behind the body of the testis. The upper part of the epididymis is larger than the lower end from which a straight tube, the vas deferens, ascends to enter the abdomen.

Accessory organs of the male

The vas deferens has thick muscular walls. It passes to the base of the bladder, where it opens into the urethra. Lying along the last part of the vas deferens on either side, between the bladder and the rectum, is the seminal vesicle. Each seminal vesicle consists of a sac about an inch and a half long. The duct leading from this sac joins the last part of vas deferens of its own side to form the ejaculatory duct.

Surrounding the ejaculatory ducts and the first part of the urethra is the prostate gland. It is about the size of a large walnut and consists of a mass of unstriated muscle fibres intermingled with racemose glands and their ducts. Its secretion passes into the urethra by several small ducts.

The prostate tends to become enlarged after middle age and constricts the neck of the bladder, and may cause retention of urine.

From the prostate the urethra curves at an angle of 90° and passes out of the pelvic cavity into the penis. This surrounds the terminal part of the urethra. The penis is an erectile organ for penetration of the vagina.

Production of spermatozoa

The epithelial cells lining the tubules of the testis bud off cells so that the tubules become full. Until puberty the tubules contain cells having

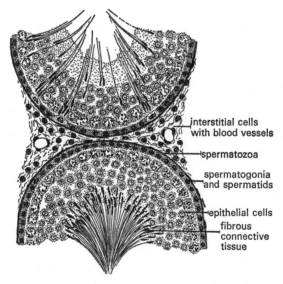

interstitial cells
with blood vessels

spermatozoa

spermatogonia
and spermatids

epithelial cells

fibrous
connective
tissue

Fig. 116. Section of two tubules of the testis (Marshal).

large nuclei. Some of these are the sperm mother cells or spermatogonia, the others protect the spermatogonia.

At puberty the spermatogonia divide up many times to form spermatocytes and then spermatids. These latter develop into the spermatozoa.

Semen, which consists of the spermatozoa suspended in an albuminous fluid, passes from the testis along the vas deferens. A secretion which probably helps to nourish the spermatozoa is passed on to the semen from the seminal vesicles. Another secretion is passed on to the semen from the prostate. This second secretion probably activates the spermatozoa, which now begin to move more quickly in the semen.

Essential organs of the female and production of ova

The two ovaries are the essential organs of the female. They are almond shaped and placed on either side of the pelvis. At birth an ovary consists of a frame of spindle-shaped cells surrounded by a layer of cubical epithelium, the germinal epithelium. The epithelium, in places, grows down into the connective tissue and forms groups of cells which become isolated from the epithelium surrounding the organ. These groups form primordial follicles.

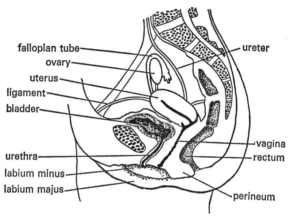

Fig. 117. Section through the female pelvis.

One cell in each follicle becomes larger than the rest and forms a primordial ovum, or primary oocyte, while the others form a single layer of flattened nucleated cells (the follicular epithelium) surrounding this ovum. In the ovary of a new-born child there are about 70,000 primordial follicles. After the onset of puberty about 500 of these mature and become ripe ova.

The follicular epithelial cells first increase in number, become cubical and arrange themselves in layers. A cavity appears in their midst and becomes full of liquid, the liquor folliculi. This liquid partially separates the epithelium into two parts, the membrana granulosa lining the follicle, and the discus proligerus around the primordial ovum. Some of the spindle-shaped cells within the ovary arrange themselves in concentric layers to form a capsule, the theca interna, around the follicle. This becomes surrounded by an outer covering of fibrous connective tissue, the theca externa.

The primordial ovum obtains its nourishment from the follicular cells.

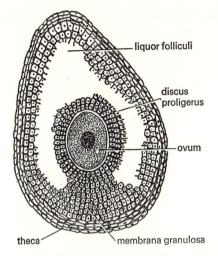

liquor folliculi

discus proligerus

ovum

theca

membrana granulosa

Fig. 118. A mature follicle before the discharge of the ovum.

It is continually increasing in size and it surrounds itself with a membrane, the zona pellucida.

As the follicle ripens in this way it passes to the surface of the ovary, the ovum divides twice, first by reduction division, then by an ordinary division producing the mature ovum and the two polar bodies, which disintegrate.

The follicular wall bursts and the mature ovum is discharged.

Corpus luteum

The cells of the membrana granulosa increase in size and almost fill the follicle. Simultaneously the cells of the theca interna divide up, and grow in among the membrana granulosa cells, completely filling the follicle. In this way the corpus luteum which produces the hormone progesterone is formed. It is nourished by blood vessels and increases in size until about the nineteenth day after the discharge of the ovum. After that, if fertilization has not taken place degeneration sets in. If the ovum has been fertilized with resulting pregnancy it continues to increase.

Accessory organs of the female

Internal. Discharged ova are conveyed away from the ovaries by the Fallopian tubes or oviducts, which are about 4 in long. Each starts with a swollen open end, the ampulla, which has a fimbriated margin. One of these fibres is attached to the ovary.

From the ampullae the oviducts pass inwards and join to form a wider tube, the uterus. The epithelium lining the oviducts is ciliated.

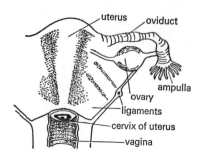

Fig. 119. The left ovary, oviduct and uterus.

The uterus or womb is a pear-shaped muscular organ situated in the pelvis anterior to the rectum and posterior to the bladder. It is about 2 to 3 in long and is divided into three parts:

(1) The fundus, which is convex. The oviducts arise from this part.

(2) The body.

(3) The cervix or neck, which communicates with the vagina, a tube leading to the exterior. The cervix is constricted.

The uterus is kept in position by ligaments and is also supported by the bladder and the muscular floor of the pelvis (levator ani).

The vagina is a distensible muscular organ lined with mucous membrane. From the vaginal orifice it passes upwards and backwards through the muscular floor of the pelvis at an angle of 45° with the horizontal. It lies in front of the rectum, and behind the bladder and urethra. At its upper end it widens to receive the neck or cervix of the uterus.

The external and internal reproductive organs of the female are separated by the hymen which is a thin fold of mucous membrane situated around the vaginal orifice.

The hymen varies in shape but normally the surfaces of the fold are in contact so that the vaginal orifice becomes a slit between them. In rare cases the hymen forms a complete septum across the lower end of the vagina, but this condition may not be discovered until after puberty when the menstrual flow cannot pass to the outside.

External. In front of the vaginal orifice is the urethral opening. The

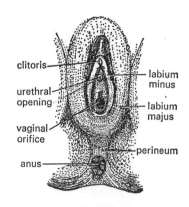

Fig. 120. External accessory reproductive organs of the female.

vaginal orifice and urethral opening are guarded by two lips of erectile tissue covered with skin. These are called the labia minora. Surrounding these small lips are two larger lips, the labia majora. In front of the urethral opening is the clitoris, an organ homologous with the penis in the male. It is very small and does not contain any part of the urethra. Lying anterior to the clitoris is a pad of fat covered with skin called the mons pubis. At puberty it becomes covered with hair. The parts around the vaginal orifice are collectively spoken of as the vulva; the area between the vulva and the anus is called the perineum.

The two oviducts open into the peritoneal cavity, and in the female infection may reach the peritoneum through the reproductive organs.

Mammary glands

The two mammary glands or breasts lie in the subcutaneous tissue in front of the chest. They are situated between the sternum and the axilla, and extend from the second to the sixth rib.

Each gland is divided into lobes supported by connective tissue. The lobes are divided into lobules. These are made up of alveoli, or small tubes lined with epithelial cells which produce milk. The small tubes open into small ducts, and these unite to form larger ducts until finally one large duct leaves each lobe. These large or lactiferous ducts pass to the centre of the breast, and swell out to form reservoirs or ampullae, where the milk is stored. The ducts from the ampullae open on the surface of

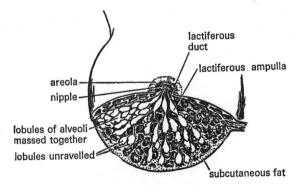

Fig. 121. Section of the right breast.

the breast at a projection called the nipple. The breasts are covered with skin. Around the nipple the skin is pigmented and forms the areola.

Mammary glands contain many lymphatic vessels. In between the lobes and just under the skin there are large fat deposits which give the breasts their rounded contour. The weight and size of the breasts varies throughout life. They enlarge at puberty and atrophy in old age. During pregnancy and after delivery, when they are actively secreting milk, they are enlarged.

In the male the mammary glands are rudimentary.

The oestrous cycle

The female genital organs undergo a cyclical change controlled by hormones every 28 days.

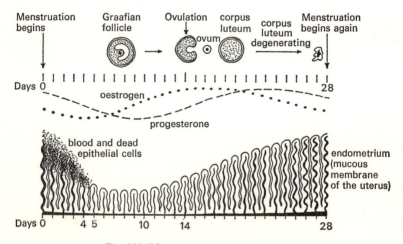

Fig. 122. Diagram of the menstrual cycle.

(1) For 5–7 days there is a discharge of blood, and epithelial cells from the uterine wall pass through the vagina to the exterior. This is known as the menstrual flow or menstruation.

(2) For seven days after the menstrual flow has ceased the uterine wall enlarges and its mucous membrane becomes regenerated.

(3) About fourteen days after the menstrual flow has begun or about 7 days after it has ceased, an ovum is discharged. The corpus luteum then develops and in the mucous membrane of the uterine wall the glands

increase in size and the blood vessels dilate. The mucous membrane becomes ready to receive the fertilized egg. This preparation is completed by the 23rd day.

(4) If fertilization has not taken place the corpus luteum degenerates

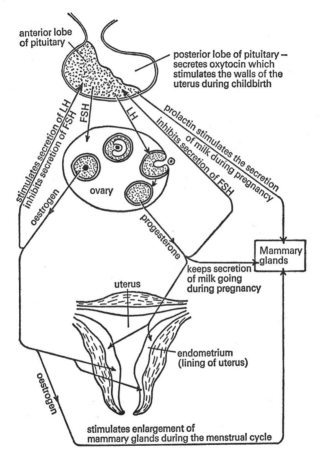

Fig. 123. The menstrual cycle and its control by hormones.

and the mucous membrane of the uterine wall breaks down. First the epithelial cells are shed and then the enlarged blood vessels burst. Menstruation begins again on the 28th day.

These cyclical changes first occur at puberty and continue throughout active reproductive life. About the age of 45–47 the ovary becomes less active; this stage of life is known as the menopause or climacteric.

Pregnancy

Fertilization if it occurs takes place in the Fallopian tube. A spermatozoon which has propelled itself up the vagina and uterus by means of its tail penetrates the ovum and a fertilized egg is formed by fusion. The fertilized egg becomes surrounded by a membrane, which prevents the entrance of any other spermatozoon. The egg then divides to form a mass of cells known as the morula, due to its mulberry-like appearance under the microscope. It has a diameter of about 2 mm.

This morula or earliest stage of the embryo passes down into the uterus. Fluid collects inside it and the cells become differentiated into an outer layer, one cell thick, known as the trophoblast, and an inner mass lying against one part of the trophoblast. The embryo is now known as the blastocyst.

The implantation of the blastocyst which takes place three to five days after fertilization is brought about by the trophoblast destroying part of the mucous membrane of the uterus.

Some cells from the inner mass spread to form a thin layer lining the trophoblast; in this way the membrane known as the chorion is formed.

The mucous membrane of the uterus which is known as the decidua, as part of it comes away when the child is born, envelops the blastocyst and fuses with the chorion.

The inner mass of the blastocyst comes to form the wall of two sacs, the amniotic cavity and the yolk sac. The cells where the two walls meet divide up, giving rise to a flat plate of cells, the embryonic disc, from which the body of the new individual, or foetus, develops. The amniotic cavity enlarges and its wall, known as the amnion, becomes fused with the chorion. The fluid which the cavity contains protects the foetus from mechanical injury. Soon the amniotic cavity surrounds the foetus, except for a stalk of tissue which attaches the foetus to the chorion. This stalk or umbilical cord conveys the blood vessels to the foetus.

The yolk sac gets smaller and gradually disappears.

The chorion sends villi into the decidua. These soon disappear except where the umbilical cord enters the chorion. At this point the villi increase in size and the chorion and decidua together form the disc-shaped placenta.

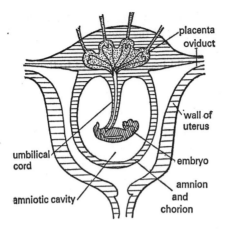

Fig. 124. Section through the uterus containing developing embryo.

The foetus grows and completely fills the uterine cavity, the fused decidua and chorion soon become fused with the rest of the decidua on the opposite side of the uterus.

The placenta enlarges and in it blood spaces develop. These spaces are filled with maternal blood.

The chorionic villi within the placenta become larger. They are full of foetal blood vessels and there is an interchange by diffusion between the foetal and maternal blood, the foetus obtaining food and oxygen and disposing of waste products. The blood in the two sets of vessels does not mix.

Parturition

The process of birth or parturition occurs nine months after fertilization. During the last few months of pregnancy movements of the foetus can be detected and its heart beats can be heard through the abdominal walls.

Parturition begins with periodic contractions of the muscles in the walls of the uterus. At the onset of labour these contractions become more pronounced and the cervical part of the uterus dilates. Pressure is exerted on the foetal membranes and they burst, allowing some of the amniotic fluid to escape.

The contraction of the uterine muscles, assisted by the abdominal muscles, expels the foetus through the vagina to the exterior.

After birth the umbilical cord is ligatured and cut, and the foetus starts its independent existence as a child. By further contraction of the uterine muscles the placenta and ruptured membranes are expelled. Under normal conditions the contraction of the muscles squeezes the blood vessels and prevents haemorrhage.

On birth when the child becomes exposed to the air it begins to respire and generally to cry.

Male and female development

After birth, a human being grows steadily until adult, but not all parts of the body grow at the same rate; the head for example is relatively larger in the baby than in the adult.

The growth rate at birth is high, but drops during the first few years. Before adolescence it is similar in boys and girls, but at ten or eleven girls shoot ahead, while in boys the spurt of adolescence usually does not begin until the age of thirteen or later. After adolescence growth takes place more slowly until full size is reached.

During adolescence changes take place in the reproductory organs. These have already been described. Other changes include, in the female:

(1) The development of the breasts.

(2) Growth of hair under the arms and on the pubic bone.

(3) Broadening of the hips, in preparation for child bearing.

In the male:

(1) Broadening of the shoulders.

(2) Growth of hair under the arm, on the pubic bone, and on the chest, face and limbs.

(3) Lengthening of the legs proportionately to those in a girl.

(4) Breaking of the voice.

The physical changes in both boys and girls are accompanied by emotional ones.

Reproductory hormones

In the male. If the testes become diseased before puberty the seminal vesicles and the prostate gland gradually disappear, and the secondary sex characters, such as the growth of hair on the chin and the deepening of the voice, do not develop.

If, on the other hand, the vasa deferentia between the testes and seminal vesicles are ligatured, the seminal tubules of the testes disappear and the interstitial cells remain. In such a condition the seminal vesicles and prostate gland are not affected and the secondary sexual characters develop normally.

It is now accepted that the interstitial cells pour directly into the blood a chemical substance, which is responsible for the development of the secondary sexual characters and the upkeep of the seminal vesicles and prostate. This chemical substance or hormone has been isolated and is called testosterone.

In the female. If the ovaries are removed before puberty the menstrual cycle never starts. If they are removed after puberty menstruation ceases and the uterus begins to break down.

During the menstrual cycle a hormone called oestrogen, given out from the follicles of the ovary, stimulates the glands in the wall of the uterus and the mammary glands. It is responsible for the development of the female secondary sexual characters.

Another hormone called progesterone which is produced by the corpus luteum is responsible for the hypertrophy of the uterus in preparation for the implantation of the embryo. If fertilization does not take place the corpus luteum degenerates; no more progesterone is secreted and the uterine wall breaks down. When an ovum is fertilized and becomes implanted on the wall of the uterus the corpus luteum increases in size and secretes more progesterone into the blood. This progesterone:

(1) Brings about a change in the uterine wall.

(2) Causes the mammary glands to increase in size.

(3) Suppresses ovulation and secretion of oestrogen during pregnancy. This causes menstruation to cease.

Removal of the corpora during pregnancy leads to abortion.

Control of reproduction by the pituitary

In both male and female the anterior lobe of the pituitary gland gives out two hormones F.S.H. (follicle stimulating hormone) and L.H. (luteinising hormone) which act on the gonads.

In the male the F.S.H. stimulates the seminal cells and thus the development of the spermatozoa, and the L.H. acts on the interstitial cells regulating the secretion of testosterone. They are secreted simultaneously.

In the female the F.S.H. which is secreted from about the fifth to the twelfth day of the menstrual cycle stimulates the growth and the maturing of the follicles and controls the production of oestrogen. The oestrogen acts on the pituitary and suppresses the secretion of F.S.H. and stimulates the secretion of L.H. The L.H. stimulates the growth of the corpus luteum and controls its production of progesterone which inhibits the secretion of F.S.H.

When the corpus luteum breaks down after an unfertilized egg has reached the uterus, the supply of progesterone passing into the blood ceases and the secretion of F.S.H. from the pituitary is no longer inhibited. In this way the cyclical activity is established.

The anterior lobe of the pituitary secretes another hormone called prolactin. This stimulates the secretion of milk which is kept going by the progesterone in the blood.

PRACTICAL

1 Examine under the microscope prepared slides of cells undergoing reduction division.
2 Examine under the microscope transverse sections of a testis, vas deferens and prostrate gland.
3 Examine under the microscope transverse sections of an ovary and mammary gland.

19 The ductless glands

Different parts of the body are co-ordinated by means of the nervous system. In addition co-ordination between organs is brought about by chemical substances which pass into the blood. These substances are called hormones or chemical messengers. The organs which produce and secrete these substances directly into the blood are called ductless glands or endocrine organs.

Some ductless glands and their secretions have been discussed; they include:

(1) The pancreas, which produces and secretes insulin (see page 137).

(2) The ovary, whose follicles produce and secrete oestrogen (see page 203), and whose corpus luteum produces and secretes progesterone.

(3) The testis, whose interstitial cells produce and secrete testosterone (see page 203).

Other ductless glands are the thyroid, parathyroids, thymus, suprarenals and pituitary.

The thyroid

This gland lies in the neck. It is divided into two lobes which lie one on either side of the trachea, and are joined by an isthmus of thyroid tissue passing in front of the trachea. Under the microscope the gland is seen to consist of closed vesicles lined by cubical epithelial cells which secrete a viscous material containing thyroxin into the cavity of each vesicle. Iodine forms part of the thyroxin molecule.

The thyroxin is continually being secreted into the blood in the capillaries which ramify through the organ.

In a young child, if the gland is undersecreting, mental and physical growth is retarded and the child becomes a cretin. Adults who lack thyroxin develop myxoedema, a disease characterized by a thickened and dry skin, baldness, slowness of speech, and slow pulse. The patient becomes fat, sluggish and mentally dull.

When a cretin or a patient suffering from myxoedema is fed on thyroid the condition improves; when on the other hand thyroid is given to a normal healthy person he becomes thin and exhausted. His condition resembles that of a person with an enlarged thyroid or goitre. Such a person, whose thyroid gland is overactive, has a very high pulse rate, and the whole body is overworking. The person is very excitable and loses weight.

An enlarged thyroid may be due to lack of iodine in the food or water. The gland, which is struggling to make thyroxin without iodine, becomes enlarged. Persons suffering from this kind of goitre sometimes have symptoms of hyperthyroidism and sometimes have symptoms of hypothyroidism. The condition is remedied by including iodine in the diet. In certain countries, e.g. Switzerland, where goitre is prevalent, an iodine salt is added to the drinking water.

When goitre is accompanied by protruding eyeballs and an excessive enlargement of the neck gland the condition is known as exophthalmic goitre, or Graves' disease.

From the above account it follows that thyroxine regulates the metabolic rate of the body, and produces normal growth and the proper functioning of all the tissues.

Parathyroid glands

These are four small glands lying two on either side of the thyroid. Their secretion controls the amount of calcium in the blood, and the laying down of calcium in bone. When they are overactive the amount of calcium in the blood increases at the expense of the bone calcium and calcium and phosphorus are excreted in the urine.

When these glands are not working properly the amount of calcium in the blood falls. This produces a condition of tetany in which certain muscles go into spasms. Unless the condition is relieved by the administration of a calcium compound via the mouth, or by hypodermal injection, the condition may prove fatal.

The thymus

The thymus is derived from two masses of tissue in the foetus which by birth have fused into a single lobed body in the mid-line in front of the trachea. Its microscopic structure is mainly that of a densely packed mass of lymphocytes.

The thymus continues to enlarge until puberty. After the twelfth year it undergoes progressive involution, by the age of twenty it is normally very small. This progressive atrophy appears to be related to the endocrine changes normally occurring during adolescence.

The only demonstrable function of the thymus is the production of lymphocytes. There is no satisfactory evidence that it is responsible for any endocrine secretion of its own in man. A large thymus sometimes persists into adult life but bears no relationship to unexplained death from shock as was at one time thought.

The adrenal or suprarenal glands

These glands lie at the upper end of the kidneys (see Fig. 86). They consist of two parts, the outer part or cortex, and the inner part or medulla.

The medulla continually produces and secretes into the blood a substance known as adrenalin.

In conditions of stress (emotion, anger, fear) the medulla is stimulated through the nervous system to overactivity and prepares the body to meet conditions which demand greater effort. The adrenalin has the same effect as stimulation of the sympathetic system. It increases the heart-rate; it causes the liver to convert its stored glycogen into glucose; it stimulates the plain muscles in the walls of the arterioles to contract and raise the blood pressure.

Adrenalin is injected into the body to counteract shock, and may be applied externally to wounds to stop bleeding.

The cortex of the adrenal gland is essential to life. If it is diseased the patient becomes wasted and progressively weaker, and his skin becomes bronzed. The condition is called Addison's disease.

It is now known that the secretion from the cortex contains several hormones. One controls the water and salt balance of the body, the disturbance of which is marked in Addison's disease; a second exerts an influence on carbohydrate metabolism; a third, cortisone, has been used in the treatment of rheumatoid arthritis. The cortical secretion must also contain some substance which influences the development of the sex glands and the secondary sexual characters.

The cortex, unlike the medulla, is not under the control of the nervous system but is stimulated to secrete by a hormone given out by the anterior lobe of the pituitary gland.

The pituitary glands

The pituitary gland lies in the skull at the base of the brain. During development the hypophysis, a pouch from the roof of the mouth of the embryo, fuses with an outgrowth from the floor of the thalamus, called the infundibulum, which becomes part of the gland. The gland is divided into two lobes, the anterior and posterior.

Anterior lobe. The anterior lobe has been called the master gland, for the hormones secreted from it control the production of secretions by various other glands. The hormones given out from the anterior lobe include:

(1) The growth hormone, which regulates growth. If over-production takes place before the epiphyses have become joined to the rest of the bone, the bones increase very rapidly in length and gigantism results. If, on the other hand, it takes place after the joining of the epiphyses, a condition of acromegaly is produced. In patients suffering from this disease, the bones of the face and limbs become enlarged.

Under-production of the growth hormone in the very young produces infantilism in which the body does not grow and the mentality remains that of a baby. Lack in older children produces dwarfism, but it does not as a rule impair mental development.

(2) The thyrotrophic hormone, which controls the production of thyroxin by the thyroid gland.

(3) The adrenocorticotrophic hormone, which controls the production of the secretions by the adrenal glands.

(4) The gonadotrophic hormones (see page 203).

(5) Prolactin which stimulates the secretion of milk after the birth of the child.

It is now recognized that there is no individual diabetogenic or pancreaticotrophic hormone produced by the pituitary. However, the anterior pituitary does exert some influence on blood sugar levels. An increase in blood sugar (hyperglycaemia) is a subsidiary action of the growth hormone. The thyrotrophic and adrenocorticotrophic hormones both tend to raise the concentration of blood sugar indirectly by increasing the secretory output of their respective target glands.

Posterior lobe. The posterior lobe gives out a secretion called pituitrin which contains two hormones, vasopressin and oxytocin, and probably others. The secretion of pituitrin causes the contraction of plain muscle fibres and a general rise of blood pressure by stimulating the walls of the blood vessels to contract. It may be used to counteract the effect of shock and to stimulate the contraction of the uterus in the later stages of labour.

Pituitrin diminishes the amount of urine excreted and increases the amount of sugar in the blood.

PRACTICAL

Examine under the microscope prepared slides of sections of the different ductless glands.

20 Inheritance

That certain characteristics are passed on from one generation to the next has been recognized from earliest times, but it was not until the nineteenth century when Gregor Mendel (1822–1884) conducted his famous breeding experiments on the common garden pea that the study of inheritance was first approached in a scientific way.

Mendel noticed that pea plants differed in a clear-cut way for several pairs of contrasting characteristics. For example, some had seeds with wrinkled seed coats and some with smooth seed coats; some had seeds with yellow cotyledons and some with green. Some plants were tall and some dwarf. In some the ripe pea pods were inflated, in others constricted between the seeds. He wondered what would be the result of crossing plants with contrasting characteristics and he set out to investigate.

Initially Mendel used lines which were breeding true, proving that this was the case by letting flowers self-pollinate themselves for several generations. He then crossed plants by pollinating the flowers breeding true for one characteristic with pollen from plants breeding true for the contrasting one and examined the results.

When for example Mendel crossed tall plants with dwarf plants the offspring, forming what is called the first filial generation F_1, were all tall and not as might have been expected of intermediate size. This result was obtained no matter which plants supplied the pollen containing the male gametes.

Mendel then self-pollinated the flowers of the F_1 generation and found that in the F_2 generation, tall and dwarf plants were produced in the ratio 3:1. In the F_3 generation all the dwarf plants bred true and so did one-third of the tall, while two-thirds of the tall plants gave a mixture of tall and dwarf plants, again in the ratio 3:1.

Mendel's results may be shown diagrammatically:

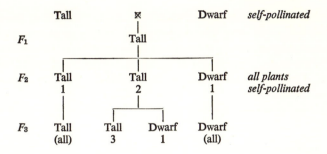

Mendel experimented with seven differences in peas in this way, at first always limiting his study to one pair of characteristics, e.g. tallness and dwarfness, at a time.

Mendel's First Law

To explain his results Mendel developed his theory. He thought that there must be factors within the plant which caused certain characteristics, such as the height of plant, to develop. He postulated that for any particular characteristic the adult possessed two factors which were not fused but separated in the formation of the gametes, the latter containing only one of the factors for a particular pair of characteristics. He also thought that the factors could be either dominant or recessive. If both the factors for alternative characteristics, e.g. tallness and dwarfness, were present, the plant would show the dominant one, tallness. The presence of the factor for the recessive character, dwarfness, would only show as a result of further breeding. Mendel's explanation may be shown diagrammatically:

Let T be the factor for **tallness** (dominant) and let t be the factor for **dwarfness** (recessive):

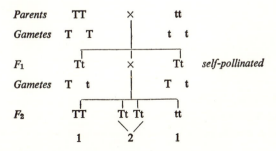

This explanation fitted Mendel's result and he proved the accuracy of it by crossing back the F_1 generation with one or other of the pure parental types.

He reasoned that if the F_1 generation were crossed with a pure dwarf plant there would be two kinds of offspring in approximately equal amounts:

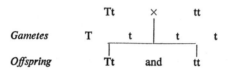

If on the other hand the F_1 generation were backcrossed with pure tall plants all offspring would be tall:

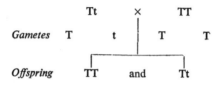

Both his predictions agreed with the experimental results he obtained and the *segregation of factors* confirmed in this way became known as Mendel's First Law.

Mendel published the result of his work in 1866. At that time no mechanism was known which could bring about the segregation of factors and his work was ignored until early in the 20th century when chromosomes were discovered and their behaviour observed (see pages 188, 189 and pages 233–235). It soon became clear that they were the physical basis of inheritance and that the small units of protoplasm called genes (see page 236) arranged along them in a linear way and occupying a definite place or locus corresponded to Mendel's factors.

A pair of genes for any particular characteristic is carried by a pair of chromosomes, one gene for a character on each chromosome. Just as the members of each pair of chromosomes separate on the formation of the gametes or germ cells, so also do the genes.

Blending inheritance

Since Mendel's time it has been shown that in many cases one of the genes is not dominant to the other. Although the two do not fuse in the fertilized egg the offspring in the F_1 generation are intermediate between the parents. For example, when a red-flowered snapdragon is crossed with a white-flowered snapdragon all the flowers of the F_1 generation are pink. On self-pollinating the F_2 generation produces plants which flower in the following ratio 1 Red 2 Pink 1 White.

Certain definitions will make the following account clearer:

Allelomorph (or allele). One of any pair of alternative characters or each of the members of a gene pair.
Homozygous. Bearing identical genes for a given character.
Heterozygous. Bearing dissimilar genes for a given character.
Phenotype. An organism classified according to its appearance.
Genotype. An organism classified according to its genetic make-up.

Mendel's Second Law: The independent association of factors (or genes)

Mendel continued his work by crossing pea plants differing from each other in two pairs of alternate characters or allelomorphs. For example, he crossed pea plants with round yellow seeds with those with wrinkled green seeds. The F_1 generation showed all round yellow seeds. When these plants were self-pollinated the types of plants produced were in the following ratio:
9 round yellow: 3 round green: 3 wrinkled yellow: 1 wrinkled green.

Mendel explained these results on the assumption that there is independent association of factors as shown diagrammatically:

Let R be the factor for Roundness.
„ r „ „ „ „ wrinkledness.
„ Y „ „ „ „ Yellowness.
„ y „ „ „ „ greenness.

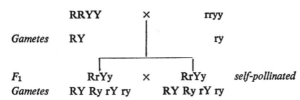

The possible combination of these gametes can be worked out by the chess board method.

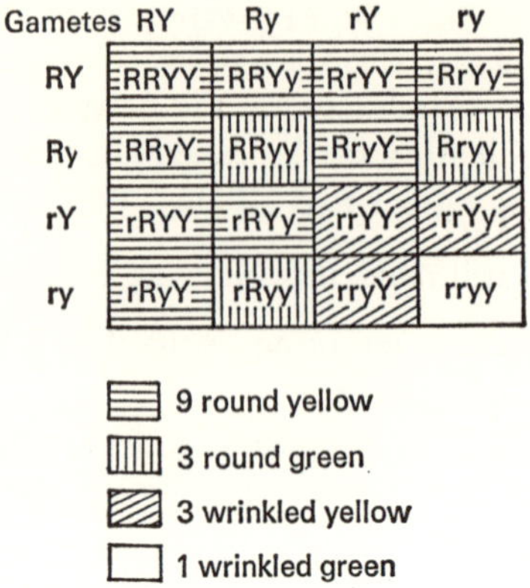

Gametes	RY	Ry	rY	ry
RY	RRYY	RRYy	RrYY	RrYy
Ry	RRyY	RRyy	RryY	Rryy
rY	rRYY	rRYy	rrYY	rrYy
ry	rRyY	rRyy	rryY	rryy

9 round yellow
3 round green
3 wrinkled yellow
1 wrinkled green

The explanation of this result based on chromosomes being the physical basis of inheritance is that the genes for the appearance of the testa and the genes for the colour of the seed are carried on different chromosomes, and are therefore transferred independently.

Linkage

Mendel's Second Law is not universally true. Sometimes two or more pairs of allelomorphs are transferred together. This is known as linkage, the genes for more than one character being linked together on the same chromosome.

For example, it has been shown to occur in Drosophila, the fruit fly, which has been used a great deal in breeding experiments as it breeds easily and fast. It has four linkage groups corresponding to four chromosome pairs.

Human heredity

In studying human heredity it is impossible to carry out controlled breeding, and biologists rely on family histories or pedigrees for their information. From these it can be deduced that characters are transferred in humans according to the Mendelian Laws.

Examples of single-gene effect in Man

(1) **Hair colour.** The red hair gene is recessive to the black hair gene. If a red-haired woman marries a black-haired man (homozygous for

black hair), all the children of this marriage will have black hair, but they will carry the 'hidden' gene for red hair.

Let B be the gene for black hair, and b, the gene for red hair:

Then   BB   ×   bb
Gametes   B   B ↓ b   b
F_1 Bb

If one of the children now takes a spouse, also with the 'hidden' gene for red hair, some of the children may have red hair.

This explains why a red-headed child may appear in a family in which the father and mother both have black hair:

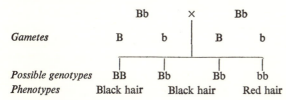

(2) **Left handedness.** Mendelian recessive.

(3) **Albinism.** Mendelian recessive.

(4) **Eye colour.** The gene for brown eyes is dominant to the gene for blue eyes. A person heterozygous for eye colour has hazel, green, or light brown eyes.

(5) **ABO blood groups.** The inheritance of blood groups follows a simple Mendelian pattern, single genes controlling the characters A, B and O.

The characters A and B are both dominant to the character O, while A and B have equal effect. There are therefore the following different types of phenotype and genotype:

Genotype	AA	BB	AO	BO	CO	AB
			gives rise to Blood Group			
Phenotype	A	B	A	B	O	AB

The inheritance of blood groups may have practical importance in paternity cases. A child might be born to an unmarried woman, who might name a certain man as father and the man might deny responsibility. Blood tests can never prove that a man is the father, but they can prove that he is not. For example, if a child belongs to blood group AB, the mother to group A, a man who also belongs to group A could not possibly be the child's father.

Hidden genes

From the above account it has been made clear that a person can carry a gene which does not show itself. Such a person is known as a carrier for that particular gene.

Sex linkage in man

Certain characters are sex-linked for they appear frequently in men, but very rarely in women. For example, red-green colour blindness affects about 4% of the males in the population but under 1% of the females. It has been shown that the genes responsible for colour vision are located on the X sex chromosome and are therefore single in the male (see page 190 for determination of sex), so that if the gene for green-red colour blindness is present, although recessive, it will produce a colour-blind male. A female, on the other hand, will only be colour blind if the two recessive genes are present.

Let X^1 be the chromosome bearing the gene for colour blindness:

	XX^1		XX	
	Carrier female		Normal male	
Gametes	X	X^1	X	Y
Possible offspring	XX	XY	X^1X	X^1Y
	Normal female	Normal male	Carrier female	Colour-blind male

A colour-blind female may only arise if a female with the hidden gene for colour blindness marries a colour-blind male:

	XX^1	×	X^1Y	
		↓		
Possible offspring	XY	XX^1	X^1X^1	X^1Y
			Colour-blind female	

In the F_1 generation there is the possibility of X^1X^1.

Other sex linked characters in humans include the disease haemophilia, but here no affected female carriers have been known.

Inheritance of skin colour

Sometimes a character is controlled by more than one pair of genes or allelomorphs. For example, skin colour is controlled by between four and eight independently segregating genes. In this case no gene is dominant or recessive, the colour of the skin of the offspring depending on the ratio of the number of genes for white skin and the number of genes for dark skin.

Let d be the gene for dark skin, and w the gene for white skin.

Suppose there are four genes for skin colour in the body cells and two in the germ cells. When a white-skinned person mates with a dark-skinned person the following represents the genetic pattern:

$$\text{dddd} \quad \times \quad \text{wwww}$$

Gametes   dd   dd | ww   ww

F_1 dd ww (all half-caste)

Suppose a half-caste then mates with a white-skinned person:

$$\text{ddww} \quad \times \quad \text{wwww}$$

Gametes   dd dw ww ww^1 ww ww

In the F_1 generation the possible genotypes would be:

ddww, dwww, wwww.

The offspring would therefore range from half and half to white-skinned.

Crossing over

During meiosis when the four chromatids of a pair of chromosomes become twisted they often break at points and interchange material, so that for example the lower part of one chromatid may join up with the upper part of another. The breaks occur regularly, but there is no way of knowing where they will occur.

Inheritance and variation

As a result of the reshuffling of genes and chromosomes and the inter-change of parts of the chromatids during meiosis variations are produced, and offspring, although resembling their parents, are not identical with them.

The particular detail of form and functioning of any phenotype depends both on the genetic make-up and on environmental influences. Man cannot radically alter the former, but he can control to a considerable degree the external and internal environment which can modify the effects of genetic characteristics. For each individual the aim must be to create an environment in which he can develop to the highest state of well-being within the limits of his constitution.

H

Appendix to Part One: The cell

CELLS = TISSUE
TISSUE = ORGANS
ORGANS = SYSTEMS .

The theme of this book could be summed up by stating that the structure of the body is suited to the functions it has to carry out. The various major tasks of the body are undertaken by the systems of organs, each of which is a 'specialist'. The organs are made up of groups of microscopic units termed cells, and materials produced by these cells. The cells are of different kinds, and groups of similar cells and their products within the organs comprise the tissues. The ways in which tissues and organs are built and behave have been quite well understood for some considerable time, and new knowledge about such matters that has been gained since this book was first published has been such as to demand only minor revision in successive editions. The position with regard to cells, however, is very different.

Structure of the cell

The 'cell theory' – the recognition that larger organisms, plant and animal, are made up of special biological units – dates from the 1830s. Since then there has been steady progress towards an understanding of the structure of cells, as far as was possible through the use of the optical microscope. The amount of detail that was recognized in this way can be seen from Fig. 125 which sums up the generally accepted features that could be seen as a result of the various methods of preparing cells in use up to the 1930s. It was not then possible to relate convincingly the structures recognized to the work done by the cell. Nowadays, though there remain many unsolved problems, we can see that this interrelation of structure and function holds at the submicroscopic level as well as for the tissues and

organs of the body. This changed situation is largely due to the appearance of new research methods in the last twenty years or so. You will
note that much of Fig. 125 is blank. The cell appeared to be filled
with a jelly-like or semi-fluid protoplasm, the structure of which was
a matter of speculation, and the consistency of protoplasm and
its chemical composition were such as to be thought colloidal in
nature.

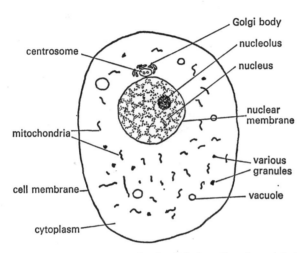

Fig. 125. A general picture of an animal cell before the advent of the electron
microscope. The protoplasm is divisible into the nucleoplasm within the nuclear
membrane, and the cytoplasm which occupies the rest of the cell.

Colloids are mixtures in which rather large molecules or molecular
aggregates are dispersed in a fluid medium, which is water in the case of
protoplasm. The individual particles of 'dispersate', each consisting of
many atoms, are large enough to have surfaces made up of numerous
atoms, surfaces which are relatively enormous compared with their
volumes. Yet the particles are too small to be seen by the optical microscope, or to fall out of 'solution' like precipitates. Matter in this colloidal
state shows great surface activity, and it was supposed that the various
complicated things that cells could do were associated with these
colloidal surfaces. There is no doubt that many of the details of cellular
activity can be accounted for in this manner, but there were many
difficulties in the way of accepting that many of the 'organized' activities
of cells could be so explained.

Modern techniques of investigation

During the last thirty years or so there have developed the methods which have given us a new insight into the working of the cell. The electron microscope has shown us fine structure in the blank areas of Fig. 125; the ultracentrifuge has enabled us to separate cell components one from another, and test their chemical capabilities; isotopes, especially radioactive isotopes, can be used to track particular atoms through the maze of changes in the cell; X-ray analysis has enabled the experts to work out the precise shape of some of the large molecules found in the cells.

Electron microscope

In preparing material for the electron microscope, the greatest care has to be taken in its handling, to avoid distortion of the delicate structures within the cells and though electron microscopes began to appear on the market just before World War II, it is only in the last decade or so that the techniques of preparation have been perfected, and that 'high resolution' electron microscopes have become available. The electron, the unit particle bearing negative electric charge, can behave as if it were a set of electromagnetic waves in some circumstances, and these waves can have wave-lengths many times – many thousands of times – smaller than those of visible light. We cannot 'see' these electron waves directly with our eyes, but they can affect suitable screens and photographic plates.

Tissues for study under the electron microscope can be 'stained' with heavy metals such as tungsten or osmium, which are attracted more to some cell components than others, so that electrons are selectively transmitted, and 'picture contrast' can be obtained. Very thin sections of the tissues or cell preparations are made, and since the molecules of gas in air would scatter electrons, the interior of the electron microscope has to be a vacuum. A full description of the procedure needed to ensure a satisfactory end product, viz. a clear photograph of the cell or part of it, would be lengthy and very technical, and is beyond the scope of this chapter.

What is important is the fact that whereas with the light microscope we are limited to effective magnifications of not more than 3,000 diameters, the electron microscope can magnify up to several hundred thousands of diameters.

Detailed structure. The structure of an animal cell as revealed by the electron microscope is shown in Fig. 126. Note how the 'blank' of areas

Fig. 125 are largely occupied by the double membranes of the endoplasmic reticulum, and how the mitochondrial granules have an internal structure,

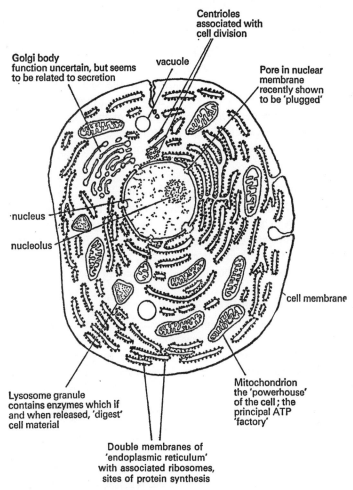

Fig. 126. Generalized diagram to show the relationship of essential cell structures as revealed by the electron microscope.

also in the form of double membranes. Many of the membranes are connected, and there are channels of communication between them and the nucleus.

The diagram is of a cell in the so-called 'resting stage', i.e. it is not undergoing cell division. The nucleus seems rather vague in its structure. Suitable techniques for 'staining' its components have not yet been developed. Even under the light microscope, a good deal is known of nuclear structure of the dividing cell. The diagram must be regarded as a kind of 'snapshot' of what is really a most active system.

Mitochondria. Careful comparison of these 'stills' of various kinds of cells shows that mitochondria are most abundant in cells which are known to be very active and may be concentrated in the parts of the cells where greatest activity is to be expected. We may suspect that the mitochondria are especially concerned with the 'energetics' of the cell.

Is it possible to separate the mitochondria and find out if they behave in a manner consistent with this supposition?

Ultracentrifuge

If preparations of cells are broken up by suitable grinding processes, the resulting 'mash' can be made to yield samples of the various ingredients by centrifuging at successively greater speeds. Heavy particles such as nuclei and unbroken cells are 'spun down' by relatively slow speeds of the centrifuge. If the fluid left is removed, and spun more rapidly, the mitochondria are then deposited and may be separated.

Activity of mitochondria

Chemical studies of mitochondrial preparations have been carried out intensively since the war. It was discovered that the 'mitochondrial fraction' was particularly good at promoting the oxidation of certain organic acids – the reactions of the Krebs cycle in fact – and was a rich source of the substance adenosine triphosphate, ATP as it is universally known, which had become recognized as a most important energy-storing material.

Since the electron microscope had revealed an internal structure for the mitochondrion, the particular part played by the inner membranes came under scrutiny. Ways were found of breaking up the mitochondria, and examining the activity of the resulting fractions. Once the mito-chondrion was broken and the fluid 'matrix' leaked out, the Krebs cycle reactions ceased, but as long as there remained the double membranes, the ATP synthesis could continue. When disruption went on to the extent that only single membranes were left, the latter ceased, but the 'electron transport' part of the overall reaction continued. Here is a precise linking of particular structures or regions with special functions,

Enzymes

✷ A SUBSTANCE PRODUCING A CHANGE

The various reactions just mentioned are able to occur by virtue of complex organic molecules – proteins – which are catalysts, ✷collectively known as enzymes. Each enzyme is very complex, but usually capable of promoting only one relatively simple reaction. There are many enzymes, so that a succession of simple reactions can mean that, overall, complex changes can be brought about.

One of the earlier difficulties in understanding cell physiology was the sheer complexity of these changes; it was hard to see how the succession of simple changes could follow one another in the proper order without some physical organization analogous to the series of assembly points alongside a conveyor belt in a factory. Recent very high powered electron micrographs of mitochondria seem to show minute granules, spaced along and attached to the mitochondrial membranes. It seems likely that each granule represents a single unit for electron transfer, bearing enzyme molecules so placed as to receive and hand over electrons. The next few years may see this story confirmed or modified, but be that as it may, it is clear that the membranes of the interior of the mitochondria represent an efficient array of varied enzymes, so that structure and function are seen to be linked at this level. Though we have as yet no detailed inform-ation about the construction of the enzymes involved in mitochondria, other enzymes are beginning to yield details of their structure to an extent which enables us at least to have some idea of the way in which they catalyse their particular reactions.

Molecular structure

To discuss this, we must make a fresh start, this time from the atomic level, and consider something of what has become known as molecular biology. Since our aim is to try to see how the three-dimensional structure of complex molecules permits chemical behaviour of biological signifi-cance, we must supplement conventional chemical formulae with diagrams of solid molecular models. The chemical elements that chiefly concern us here are hydrogen, carbon, oxygen and nitrogen, and we shall have occasion to refer also to sulphur and phosphorus.

The **hydrogen atom** is made up of a single heavy proton, bearing positive electric charge, around which moves a single light electron (about 2,000 times less massive than the proton) occupying a 'shell' or cloud which in effect constitutes the outer limit of the atom. This shell can hold two electrons, and in the helium atom, which has a nucleus of two protons and two neutrons (of a mass similar to protons, but with no electric charge), the shell is fully occupied by two electrons.

In the **carbon atom,** there are six proton and six neutrons in the nucleus (accounting for the atomic weight 12), an inner (complete) shell of two electrons, and an outer shell of four electrons. There can be up to eight electrons in this second shell, so there are four 'electron vacancies' in the carbon atoms, and electrons can be contributed by other atoms (by sharing) to the carbon atom to fill these vacancies. The carbon atom can accommodate four hydrogen atoms in this way (each joined by a single bond). The electrical forces involved determine the form of the methane (CH_4) molecule so formed – the four hydrogen atoms occupy sites on the carbon atom at the four corners of a tetrahedron:

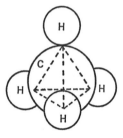

The **nitrogen atom** has seven protons and seven neutrons in the nucleus (atomic weight 14), a shell of two electrons and a shell of five electrons. The compound of nitrogen and hydrogen, ammonia (NH_3) has the hydrogen atoms occupying three tetrahedral corners:

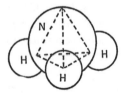

In the **oxygen atom** (atomic weight 16) there are eight protons and eight neutrons in the nucleus, a shell of two electrons, and a shell of six electrons. (Note that in all cases the number of protons is the same as the number of electrons, i.e. the atom is as a whole electrically neutral.) In water (H_2O) two hydrogen atoms contribute their electrons to complete the shell of eight; these atoms are placed on two tetrahedral corners:

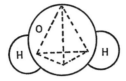

Oxygen has two electron vacancies per atom, so two oxygen atoms together can accept the four electrons of a carbon atom. 'Double bonds' are formed in this case, so that tetrahedral edges rather than corners are involved, and a molecular model of carbon dioxide (CO_2) has the following appearance:

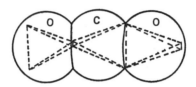

Fig. 127 illustrates some other biologically important molecules. You will observe that combination by electron sharing readily occurs between carbon atoms themselves.

Macromolecules

Carbohydrates

Some of the biologically most important molecules are very large indeed compared with the hydrogen atom, but these are built up from simpler units, linked together in a consistent way. These 'building units' include glucose ($C_6H_{12}O_6$) and ribose ($C_5H_{10}O_5$) amongst the carbohydrates; these formulae simply give the numbers of the atoms, and give no hint of their arrangement. If a 'hydroxyl group' ($-OH$) from one glucose molecule combines with a hydrogen atom (H) from a second glucose molecule (to form water) the remnants or 'residues' of the two molecules unite to form maltose:

$$2C_6H_{12}O_6 \longrightarrow C_{12}H_{22}O_{11} + H_2O$$

If this *condensation* process continues, very large molecules of starch, or of glycogen, which are polysaccharides, may be formed. (Glucose and ribose are monosaccharides, maltose is a disaccharide.) In digestion the reverse process occurs, and the large molecule breaks down into the

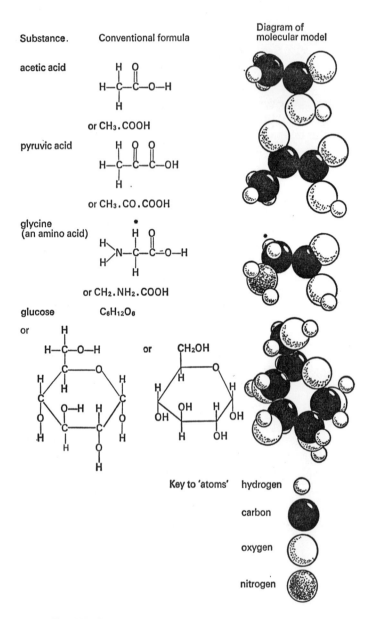

Fig. 127. Some biologically important organic molecules.

monosaccharide units. Digestion, involving breakdown by the chemical action of water, is an example of hydrolysis.

$$(C_6H_{16}O_5)_n + nH_2O \quad \underset{\text{condensation}}{\overset{\text{hydrolysis}}{\rightleftharpoons}} \quad nC_6H_{12}O_6$$

(n may be several hundreds).

Fats

Fatty acids are units which have long 'hydrocarbon' chains, each with a 'carboxyl' acid group at one end. Petrol and kerosine are hydrocarbons, so it is not surprising to find that fats, which are condensation products of fatty acids with glycerol ($C_3H_8O_3$), are good fuels. Fatty acid molecules are large for 'units', but since only three such molecules are condensed with one glycerol molecule to form a fat molecule, the latter is quite small as condensation products go. Because the fatty acid molecules are long, it is more convenient to give the formula in symbol form. Each line or 'bond' in the formula implies a pair of shared electrons. The chain molecule of stearic acid, shown below, is really linear but has been 'folded'.

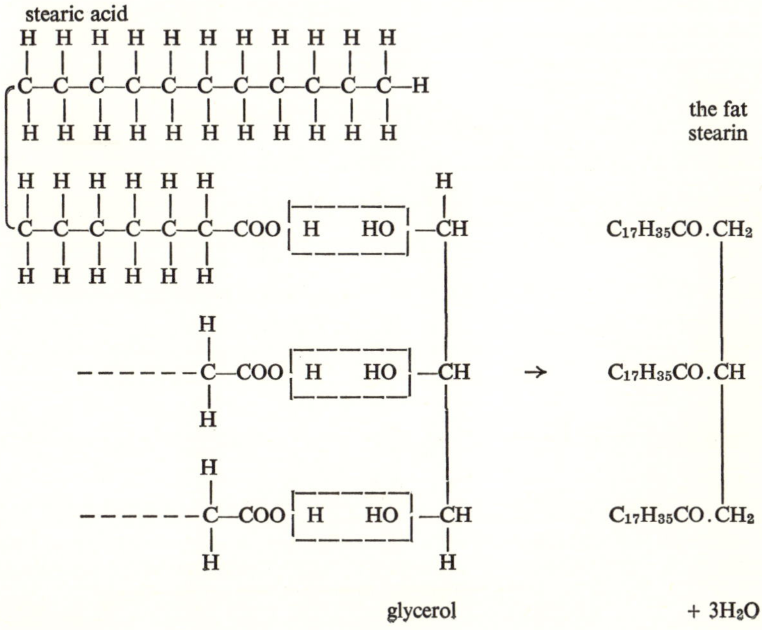

the fat
stearin

$C_{17}H_{35}CO \cdot CH_2$

$C_{17}H_{35}CO \cdot CH$

$C_{17}H_{35}CO \cdot CH_2$

glycerol

$+ 3H_2O$

Amino acids and proteins

There are about twenty of the amino acids which rank amongst the most important of biological building units. The glycine molecule has already been illustrated (Fig. 127); in the remaining amino acids, one of the hydrogen atoms is replaced by a group of atoms. These groups are listed in Table 1. Note the variety of these groups (also referred to as radicals or R groups), in structure and chemical nature.

Table 1. THE PROTEIN-FORMING AMINO-ACIDS: $R-\overset{\displaystyle H}{\underset{\displaystyle NH_2}{C}}-COOH$

Name	*Formula of R-group*	*Characteristic of R-group*[1]
glycine	H—	hydrogen, neutral
alanine	CH_3-	⎫
leucine	CH_3 　　＼ 　　　$CH-CH_2-$ 　　／ CH_3	⎪
isoleucine	CH_3-CH_2 　　　　＼ 　　　　　$CH-$ 　　　　／ 　　CH_3	⎬ hydrocarbon, neutral
valine	CH_3 　　＼ 　　　$CH-$ 　　／ CH_3	⎭
phenylalanine	HC$\overset{\overset{\displaystyle H\ H}{\displaystyle C=C}}{\underset{\underset{\displaystyle H\ H}{\displaystyle C-C}}{\big\langle \ \ \big\rangle}}C-CH_2-$	aromatic, neutral
serine	CH_2OH-	primary alcohol, neutral
threonine	$CH_3-CHOH-$	secondary alcohol, neutral

tyrosine

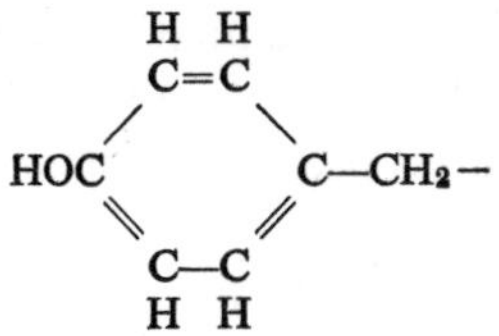

$$\text{HOC} \overset{\displaystyle \overset{H\ H}{C=C}}{\underset{\displaystyle \underset{H\ H}{C-C}}{}} C-CH_2-$$

aromatic (phenolic), neutral

tryptophan

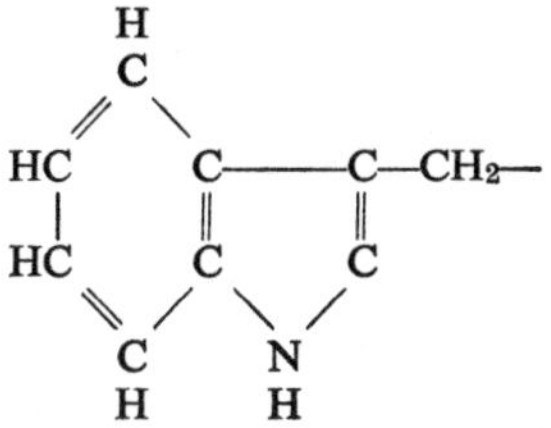

$$\overset{\displaystyle \overset{H}{C}}{HC} \qquad C \text{------} C-CH_2-$$
$$HC \qquad C \qquad C$$
$$\underset{H}{C} \qquad \underset{H}{N}$$

aromatic (heterocyclic), neutral

cysteine $SH-CH_2-$ thioalcohol, neutral

methionine $CH_3-S-CH_2-CH_2-$ thioether, neutral

asparagine $H_2N-CO-CH_2-$

glutamine $H_2N-CO-CH_2-CH_2-$

}amide, neutral

aspartic acid $COOH-CH_2-$

glutamic acid $COOH-CH_2-CH_2-$

}carboxylic, acidic

lysine $NH_2-CH_2-CH_2-CH_2-CH_2-$ amino, basic

arginine $NH_2-\overset{\displaystyle \underset{\displaystyle \underset{H}{N}}{\|}}{C}-NH-CH_2-CH_2-CH_2-$ guanido, basic

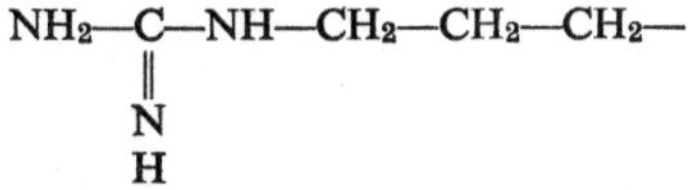

histidine

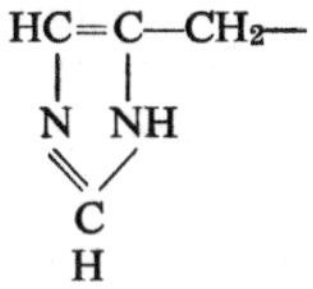

$$HC=C-CH_2-$$
$$\underset{N}{|} \quad \underset{NH}{|}$$
$$\underset{H}{C}$$

imidazole (aromatic heterocyclic), basic

proline

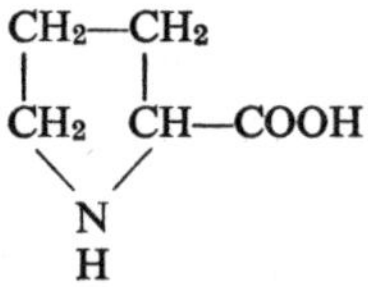

$$\begin{array}{cc} CH_2-CH_2 \\ | \qquad | \\ CH_2 \quad CH-COOH \\ \diagdown \quad \diagup \\ N \\ H \end{array}$$

This is the complete formula of what is strictly an 'imino' acid, containing a secondary amino group. It behaves like an amino acid, and occurs in *all* proteins.

[1] For the meaning and significance of unfamiliar terms, see books on organic chemistry and/or biochemistry.

In the formation of larger molecules from amino acids, condensation occurs between the acid ($-COOH$) group of one and the amino group ($-NH_2$) of another. The amino group is basic (cf. ammonia).

$$
\begin{array}{cc}
\text{amino acid 1} \left\{ \begin{array}{c}
\mathrm{H} \quad \mathrm{H}^a \\
\diagdown \diagup \\
\mathrm{N} \\
| \\
\mathrm{R_1}{-}\mathrm{C}{-}\mathrm{H} \\
| \\
\mathrm{C}{=}\mathrm{O} \\
| \\
\mathrm{O} \\
| \\
\mathrm{H}
\end{array} \right.
&
\begin{array}{c}
\mathrm{H} \quad \mathrm{H}^a \\
\diagdown \diagup \\
\mathrm{N} \\
| \\
\mathrm{R_1}{-}\mathrm{C}{-}\mathrm{H} \\
| \\
\mathrm{C}{=}\mathrm{O} \\
|
\end{array}
\\[1em]
\longrightarrow \mathrm{H_2O} \; + &
\\[1em]
\text{amino acid 2} \left\{ \begin{array}{c}
\mathrm{H} \quad\;\; \mathrm{H} \\
\diagdown \diagup \\
\mathrm{N} \\
| \\
\mathrm{R_2}{-}\mathrm{C}{-}\mathrm{H} \\
| \\
\mathrm{C}{=}\mathrm{O} \\
| \\
\mathrm{O} \\
| \\
\mathrm{H}^b
\end{array} \right.
&
\begin{array}{c}
\mathrm{H}{-}\!\!-\mathrm{N} \\
| \\
\mathrm{R_2}{-}\mathrm{C}{-}\mathrm{H} \\
| \\
\mathrm{C}{=}\mathrm{O} \\
| \\
\mathrm{O} \\
| \\
\mathrm{H}^b
\end{array}
\end{array}
$$

The 'double' molecule on the right is a _dipeptide;_ by further condensation of amino acids (acid group with NH_2 group *a*, amino group with COOH group *b*, etc.) long chains of _polypeptides_ are formed. Once the active COOH and NH_2 groups are 'neutralized' in this way, the properties of such polypeptide chains depend very largely on the particular R-groups incorporated, and their linear arrangement. The chains may be bent, folded, wound into helices, or cross-linked (especially through sulphur-containing R-groups), the particular form of the resulting protein depending, it seems, on the way in which R-groups interact with one another and on conditions in their surroundings.

By methods involving the use of **X-rays,** the actual three dimensional forms of several proteins have been determined. By chemical analysis, the sequence of amino acids in other proteins including some enzymes has been discovered.

Lysozyme

In 1965, both amino acid sequence and spatial arrangement were worked out for the enzyme lysozyme, which is found in tears and saliva and promotes the breakdown of bacterial cell walls. A model of this molecule

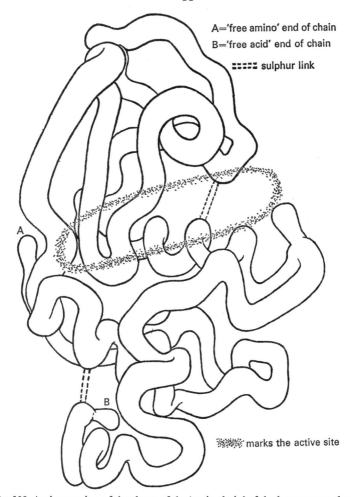

A='free amino' end of chain
B='free acid' end of chain

:::::: sulphur link

marks the active site

Fig. 128. An impression of the shape of the 'main chain' of the lysozyme molecule.

looks like an untidy tangle (Fig. 128), far removed from the regularities revealed by the electron microscope. But it seems that, in a kind of groove on one side of the molecule, R-groups occur, of such a kind and so spaced as to 'fit' part of the bacterial cell wall material, and with the aid of a couple of oxygen atoms, to break it at one point. So there is also parity of structure and function at molecular level in enzymes. This has long been suspected, but only very recently actually demonstrated in a particular instance.

Nucleic acids and enzyme synthesis

Now, these enzymes are manufactured 'on the spot' as needed, in the ordinary course of events. How are they assembled from their constituent amino acids? Only very recently has some sort of answer to this question become a possibility. To attempt it we must consider some more 'building units'. Adenine, cytosine, guanine, thymine and uracil are the names given to a biologically important series of nitrogenous bases. They are somewhat complex substances, and the formulae are worth recording only for reference, and as a reminder that nitrogen can confer basic properties. For most purposes that concern us, we may simply refer to them by their initials—A, C, G, T, U.

Table 2. THE NITROGENOUS BASES OF THE NUCLEIC ACIDS

(1) Pyrimidines (single ring of C and N atoms):

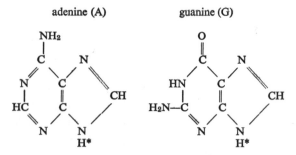

(2) Purines (double rings of C and N atoms):

If molecular models are made of these, using the 'tetrahedral rule' discussed earlier, it will be seen that the molecules form rather flat 'sheets'.

* Point of linkage with ribose or deoxyribose.

Adenine may be condensed with ribose to form the compound adenosine; to this may be attached a phosphate group, forming adenosine monophosphate (AMP). To the first phosphate group so introduced may be added a second, to give adenosine diphosphate (ADP), and to this second phosphate group may be attached a third. The substance so formed – adenosine triphosphate – is the ATP mentioned earlier. The attachment of the second and third phosphate groups demands rather a large amount of energy, which is released when the phosphate groups are removed. It is in the form of these high energy phosphate bonds that energy is stored for instant use in the cells.

Nucleotides

By combining adenine, ribose and a phosphate group in a different way, a type of compound known as a nucleotide is formed. The other nitrogenous bases also form nucleotides, and these may join end to end, through the ribose and phosphate groups, to form elongated condensation products, with the bases as side-groups. These long polynucleotide molecules are nucleic acids.

Nucleic acids

There are two main kinds of nucleic acids. In one kind, the ribose sugar is normal, and the acid is ribonucleic acid, or RNA; in the other, the ribose lacks one of the oxygen atoms, and is known as deoxyribonucleic acid or DNA. It can be shown by the use of specific dyes that the DNA is found in cell nuclei, in the chromosomes which are clearly seen in cell division. The nitrogenous base U is not found in DNA. RNA is found in the nucleolus of the nucleus, in the ribosomes attached to the endoplasmic reticulum, and in circumstances which will be discussed shortly. In RNA, U replaces the T of DNA.

Structure of DNA

Advanced studies on the nucleic acids, especially those involving X-ray methods, have shown that the DNA has a remarkable helical, double-stranded structure in which two long polynucleotide chains complement one another, in a manner which depends on the shape and electrical properties of the nitrogenous base molecules. A will pair up with T, and C with G, to give similar flat double molecules, linked by so-called hydrogen bonds (Fig. 129).

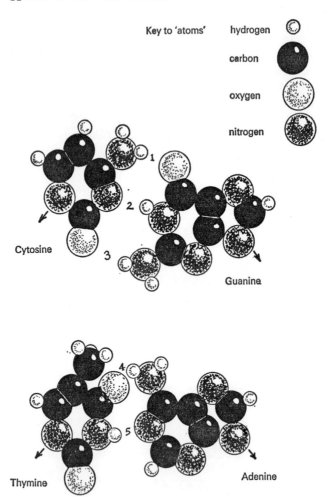

Fig. 129. 'Base pairing' in nucleic acid.
In these diagrams of molecular models, corresponding bases are arranged to show how 'hydrogen bonds' (1, 2, 3, 4, 5) may be formed when 'exposed' hydrogen atoms are able closely to approach 'exposed' oxygen or nitrogen atoms. (Atoms from which arrows extend are linked with sugar of nucleotide.)

Replication. If the two strands of DNA can be parted, and a supply of deoxyribonucleotides, ATP (for energy) and the appropriate enzymes are present, an exact copy of the partner strand will be set up on the original strand as template by this hydrogen bonding, and since each strand thus generates the missing partner, the effect is exactly that of making two identical strands, or of 'replicating' the DNA. This is the process which precedes cell division, and makes it possible for exact copies of the original chromosomes to be passed into daughter cells.

Relationship between DNA and RNA

But DNA does not merely possess this extraordinary property of self duplication. Under appropriate conditions in the undividing cell, copies of a single DNA strand may be made with ribonucleotides, to make RNA, so that whatever the order of bases in the DNA strand, it will be matched by the RNA copy. It has been shown that this RNA copy – *messenger* RNA – moves from the nucleus to the ribosomes which a number of studies have shown to be associated with protein synthesis.

Messenger RNA and polypeptide. What is the link between them? This question, hinted at in the last few sentences, and suggested by a great deal of evidence from work on bacteria and the special viruses that 'parasitize' them, seems well on the way to being answered.

Within polypeptides there are twenty amino acids to be considered. Within the nucleic acids, differences somehow involving the four bases in each kind must be involved. Is it possible that some kind of 'code' of just four 'characters' can register twenty different amino acids? If each base 'specified' one amino acid, only four of the latter could be manipulated. Two bases could be combined together, from the set of four, in sixteen different ways, so a code of two bases is also insufficient for our purpose. Three bases from the set of four can be selected in sixty-four different ways, however, which permits of several codings for each amino acid.

It does now seem quite well established that when in position on a ribosome, each set of three bases in the messenger RNA in turn 'calls' the appropriate amino acid into position and condenses it with the one that had preceded it, and by what is in effect a 'reading off' of the bases in threes in this way, the proper polypeptide is produced.

There is a further type of RNA concerned in this process. *Transfer* RNA picks up the amino acids, by a process involving ATP and enzymes, and makes them ready for this assembly process.

Now, the instructions for making the RNA came by direct copy from the DNA, so that in the chromosomes of the nuclei of each cell there are instructions ready to be followed when the right circumstances arise.

Research is now actively being carried out into the way in which particular sections of DNA are brought into operation at the right time. The unit sections of DNA are the genes one encounters in the study of heredity. The matters just discussed have been summed up in the phrase 'one gene, one enzyme'.

The cell in its environment

The cell in the body may be regarded as a kind of self-maintaining system. Given a supply of suitable materials from without, including water, a selection of amino acids, oxygen, glucose, some vitamins and a variety of minerals including phosphates, and granted a 'refuse disposal service' as it were, so that carbon dioxide and excess materials and end and waste products can be removed, a cell can look after itself, grow, divide, and carry out its special function. Indeed, from one point of view, the purpose of the body as a whole is to maintain the conditions in which the cell can function properly.

Formation of ATP

It remains to say a little more about the complex of processes that go on in the cell – the metabolic changes which are often interlinked and include the building of some units such as ribose and the nitrogenous bases. To go into great detail would require a textbook of biochemistry, so we may take the problem of ATP formation as an example, especially as we have already said a good deal about the mitochondria with which it is so closely involved.

To make ATP from ADP and phosphoric acid requires at least 8,000 calories per gram molecule. When a gram molecule of hydrogen is oxidized to form water (vapour) it produces about 57,000 calories. When this latter process occurs in one step, it is a vigorous, even explosive reaction, and molecular hydrogen is not commonly found in cells! Yet it is from hydrogen that the energy for building ATP is mainly derived. The hydrogen used is mainly that carried by the glucose ($C_6H_{12}O_6$) molecule. Glucose molecules have chemical energy stored in them, energy which ultimately comes from the light which permitted photosynthesis in a green plant.

A cell presented with glucose can be thought of as having two problems, to remove the hydrogen in convenient form for oxidation, and to dispose of the 'carrier' carbon.

A glucose molecule is first made into glucose phosphate – an ATP molecule has to be 'borrowed' for this – and by a succession of changes each controlled by an enzyme, has a second phosphate group added from a second ATP molecule, and is broken into two *triose phosphate* molecules (the equivalent of $C_3H_6O_3$ plus phosphate). Each of these triose phosphates takes up a further phosphate group – directly from phosphoric acid this time – to make *triose diphosphate.*

Further changes that occur include the removal of two hydrogen atoms from each of the triose units. These hydrogen atoms are taken into combination with a *coenzyme,* and there held to be handed over when necessary. (It is of interest to note that some of the component chemicals which go to make up such coenzymes, include substances which are among the B group of vitamins. Only some of the many enzymes require such accessory substances, and the enzymes themselves are only part of the body's protein, so that compared with the protein in our food, from the amino acids of which our cells build human proteins, we need only very small quantities of such accessory substances as the cells cannot themselves manufacture.)

Chart 1. THE FORMATION OF PYRUVIC ACID

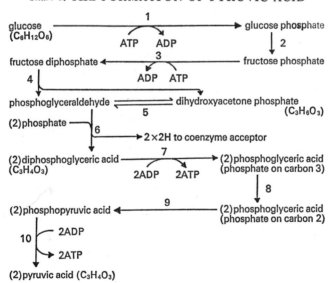

Stages 1 to 10 are controlled by enzymes. The formulae in brackets indicate the composition of the compound, *less phosphate* where appropriate.

Further enzymes control 'molecular rearrangements' in what is left of the two triose phosphate molecules, and the four phosphate groups that they bear are set in 'high energy' situations, so that four ATP molecules may now be synthesized – giving a net gain so far of two ATP molecules.

The $C_3H_4O_3$ molecules which are left from the reaction undergo further rearrangement, until two molecules of *pyruvic acid* (mentioned and illustrated earlier) are formed: $CH_3.CO.COOH$.

These changes seem to occur in the general cytoplasm, but it would not be surprising if further research revealed some underlying spatial organization. They are set out in a little more detail in Chart 1.

Krebs cycle

The rest of the 'handling' occurs in the mitochondria. The pyruvic acid is 'oxidatively decarboxylated', i.e. CO_2 is removed and O added, and the '2-carbon' compound acetic acid ($CH_3.COOH$) is formed. With the formation of carbon dioxide, the first of the three carbon atoms in each of the two triose units is 'disposed of'. (The acetic acid is at this stage

Chart 2. THE PROCESSING OF PYRUVIC ACID

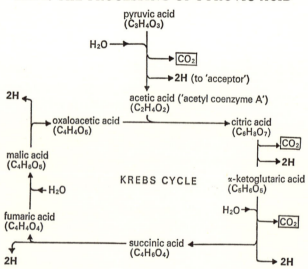

Note (a) $3CO_2$ is given out for $C_3H_4O_3$ taken in.

(b) 5 sets of 2H produced and handed to 'acceptors'. Some of the hydrogen comes from H_2O taken, the oxygen being retained in the cycle for the time being, but eventually lost in CO_2 'to keep the balance'.

present in the complex form of acetyl coenzyme A, a substance worth mentioning because in its make-up another B-group vitamin occurs, and because it can be used as the starting point of a whole series of syntheses, including fatty acids and amino acids.)

The 2-carbon acetic acid is combined directly with an acid containing 4 carbon atoms (oxaloacetic), and by a series of enzyme-controlled reactions, set out in more detail in Chart 2, the 6-carbon citric acid so formed undergoes a series of reactions whereby hydrogen atoms are removed and made ready for oxidation, two further carbon dioxide molecules are released, and the original 4-carbon acid is regenerated and able to accept another acetic acid unit.

This cycle of reactions is known as the tricarboxylic acid cycle, or Krebs cycle, after Sir Hans Krebs, who worked out the essentials of the series. This was in fact done without using radioactive isotopes, but confirmation and extension of this kind of biochemical knowledge owes much to their use.

The **hydrogen atoms,** now held by their carrier substances, come under the control of the enzymes previously referred to on the inner mito-chondrial membranes, and are handed down a series of acceptors, losing energy at each step. At one point they hand over their electrons to cytochromes, which are proteins containing iron atoms, and themselves are set free as hydrogen ions (i.e. each now carries a positive electrical charge). The electrons pass along a series of cytochromes, and are finally handed to oxygen atoms – one pair of electrons to each oxygen atom. Two hydrogen ions then combine with this negatively charged oxygen atom to make a water molecule.

Two or three of the hydrogen or electron transfer stages involve energy drops of more than 8,000 calories per gram molecule, and at each of these ADP can be combined with free phosphate to form ATP.

Many of the substances that come into these stepwise and cyclic reactions may take part in others of like or greater complexity, and we may in principle construct a mental picture of the cell, carrying out balanced, interlocked orderly processes, controlled by enzymes it itself produces, making and modifying the very framework on and in which these changes occur.

Though we can say that the fitness of structure for function is appropriate at sub-cellular level, the study of detail has barely begun. Yet, although the topics discussed in this chapter nearly all belong not only to this century, but to the last few decades, they are very old-fashioned biology indeed. The cells of the early organisms on earth, hundreds of millions of years ago, must have worked very much in the same way. In the rest of the book, attention is concentrated on the way in which one of the most recent species – *Homo sapiens* – one of the most up-to-date models, so to speak, has a structure and function which serves and is served by his constituent cells.

FURTHER READING

The Appendix attempts to do no more than introduce the 'molecular biology' which underlies the working of the body. Students who wish to explore further this new area of interest may find the following reading list useful.

Up-to-date and comprehensive discussions of cell biology may be found in: *Cells and Cell Structure* by E. L. Mercer (Hutchinson); *The Cell* by C. P. Swanson (Prentice-Hall); and *Cell Structure and Function* by A. G. Loewy and P. Siekevitz (Holt, Rinehart and Wilson).

The first of these is British, the other two are American. All are available in paperback form.

The Microstructure of Cells by Stephen W. Hurry (John Murray) contains a splendid selection of electron micrographs, with explanatory text.

The fundamental chemistry for the understanding of molecular biology is well set out in *The Genetic Code* by Isaac Asimov (John Murray).

The structure of lysozyme is considered in detail in the November 1966 issues of both *Science Journal* and *Scientific American*.

B.B.C. pamphlet entitled *What is Life?*

Part two

Hygiene

1 Personal hygiene

The health of a nation depends on the health of the individual. The State provides various health services, but it is the duty of individuals to keep themselves fit by observing simple rules of health, with regard to cleanliness, habits, diet, suitable clothing, exercise and recreation, and rest.

Cleanliness

Of the skin

The body gets rid of certain waste products in the sweat and the sebaceous glands give off grease. If this organic matter is allowed to accumulate, it produces an unpleasant smell and a medium for the growth of germs; it clogs up the sweat ducts and the sweat glands cannot function properly. To remove it, the skin should be washed thoroughly with soap, which emulsifies the grease. If possible the body should be bathed or sponged every day and parts of the body which perspire a great deal should be sponged more often. Very hot baths (38–41°C) should only be taken before retiring, if at all, since they are exhausting. Hot baths during the day should be followed by a cold sponge; warm baths (32–37°C), which can be taken with safety at any time, have a sedative effect. Cold baths, if they suit the individual, are excellent first thing in the morning, since they stimulate all the vital functions of the body. Friction with a towel is valuable in stimulating circulation.

Hands and nails should be washed thoroughly before meals; the nails should be cleansed with a wooden instrument, and kept short to prevent them from harbouring germs. Toe-nails should receive as much attention as finger-nails.

Of the head

The hair should be brushed to prevent it becoming matted and should be washed frequently. Great care should be taken, especially in children, to prevent the harbouring of parasites (see page 323).

Of the mouth and teeth

An unhealthy condition of the mouth and teeth can affect all parts of the body, producing much ill-health.

Teeth should be brushed at least twice a day, after every meal if possible, with a moderately hard toothbrush. The toothbrush should be moved backwards and forwards across the teeth, and then away from the gums in front and behind the teeth to dislodge any particles of food.

If food remains it is decomposed by bacteria; the acids produced destroy the enamel and expose the dentine to the action of bacteria. The bacteria can then penetrate the pulp cavity causing pain (see page 118). The ease with which a tooth decays depends on the hardness of the enamel. This varies from person to person, but everyone should pay frequent visits to the dentist to allow him to stop any tooth which has begun to decay.

In recent years it has been shown that the addition of small quantities of fluorine compounds to the drinking water supply very markedly increases the resistance of the enamel to decay.

The power of the teeth to resist decay depends on the general health of the person and on his diet. The diet should contain calcium, and vitamins A and D. Vitamin D and calcium are necessary for the laying down of the enamel and dentine, while vitamin A promotes healthy gums.

A good flow of saliva removes food particles lodged between the teeth. This flow is reduced during mouth breathing, and also at night. The teeth should be cleaned before retiring.

Infants should be encouraged to exercise their jaws with hard biscuits such as rusks; this stimulates the growth of the jaw to a large enough size to hold all the teeth.

Good habits

The cultivation of good habits in children is important. The bowels should be emptied regularly every day to prevent constipation. Any tendency to this may be overcome by regular exercise and a diet containing roughage and fluids.

Children should be encouraged to hold themselves well and to breathe deeply. A good posture and correct breathing stimulate the proper functioning of all the organs and help to maintain health.

Diet

A good diet is important in maintaining health, and this subject is discussed in a separate chapter. Here, however, it may be emphasized that meals should be eaten slowly, at regular hours, and the food should be properly masticated. If these simple rules are not observed, many digestive troubles such as indigestion, dyspepsia, which lead on to more serious disorders, may occur.

Clothing

Clothes are worn for several reasons. They protect the body from injury, from weather changes, and from too great loss of heat. They are also worn for decency and decoration.

Heat is lost from the body by conduction, radiation, convection and the evaporation of sweat. There is a layer of air between the skin and the clothes, and the heat is radiated across this to the inside surface of the first garment. It passes through the clothes to the outside air. The rate at which heat is conducted away from the body depends not only on the temperature of the outside air, but also on the materials of the clothes. Air is a bad conductor of heat, and any material, e.g. wool, which traps air between its meshes, is a bad conductor.

The body is continuously giving off sweat, which evaporates immediately into the layer of air surrounding the body. The water vapour diffuses through the air entangled in the clothes to the outside. If sweating is excessive and drops of perspiration form, they are absorbed by the clothing. This water passes through the clothes to the outside air, and if the latter be warm and dry the water soon evaporates; if the day be cold and damp the water remains in the clothes, making them feel wet and clammy.

The rate at which sweat is removed from the body depends not only on the amount of water vapour in the air, but also on the materials of the clothes. Some materials, e.g. cotton, absorb water readily and give it up readily. Other materials, e.g. wool, absorb water slowly, and give it up slowly. Damp cotton feels far colder than damp wool. Cotton absorbs water more readily than linen, linen than silk, and silk than wool. They give up water in the same order.

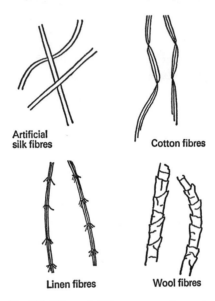

Artificial
silk fibres

Cotton fibres

Linen fibres

Wool fibres

Fig. 130. Fibres of different types of material.

Different types of material

Materials or fabrics for clothes are made from fibres which may be natural in origin, such as wool, silk, cotton and linen, or man-made, e.g. rayon, acetate rayon, nylon and terylene.

Cotton is made from the white downy fibres attached to the seeds of the cotton plant. These fibres are flat and twisted, and trap very little air. Cotton is therefore a good conductor of heat. It absorbs moisture quickly and dries quickly. For these reasons it is a good material for warm weather clothes and sports clothes.

Nurses' uniforms are made from cotton as it is easily laundered, does not readily pick up dust and germs, and is not damaged by ordinary disinfectants. It is cheaper than wool.

If cotton is woven into cellular material it has the advantages of wool without the disadvantages. Cellular material is therefore good for underwear.

Flannelette is cotton material, but as its surface is roughened it is a bad conductor of heat. It is inflammable and legislation is such that it must be made flame-resistant before leaving the factory.

Wool is made from the hair of sheep and consists of tubular fibres with overlapping scales on the outside. It is covered with a greasy substance and therefore takes up water slowly. On the other hand, it can retain a relatively large volume of water and give it up slowly in drying. Since it traps much air between its meshes it is a poor conductor of heat.

As a result of all these factors wool is a good material for clothing in cold weather. It conducts heat away from the body slowly and does not allow a too rapid evaporation of sweat. Since it has a good capacity for retaining water it absorbs all the sweat from the surface of the body.

Linen is made from the stems of the flax plant, by steeping them in water until the fibres separate.

Linen is a better conductor of heat than cotton, and can only hold a small amount of moisture. It is a bad material for underclothes, but since it launders well and does not readily pick up dirt and germs, it is useful for outer garments in hot weather. It is more expensive than cotton.

Silk is made from the threads covering the silkworm cocoon. It is not such a good conductor of heat as cotton, but a better conductor than wool. It absorbs water slowly, and cannot hold much, soon becoming sodden. For this reason it is not as good as wool for underclothing, although it does not shrink or irritate the skin. It is more expensive than wool.

Man-made fibres. Certain materials, e.g. rayons, also known as artificial silks, are made from filaments which are manufactured by forcing through a spinneret a liquid produced by the action of chemicals on wood pulp or cotton fibres. They are cheaper than pure silk, but inferior for underclothes since they are better conductors of heat and absorb less water. If the filaments are woven into a network, the quality of the material is improved.

The starting points for filaments of the new synthetic materials such as nylon and terylene are chemicals found in coal tar. These are condensed together to form a thick liquid which is then formed into filaments by forcing it through a spinneret in the same way as in the manufacture of filaments for other man-made materials.

Synthetic materials are very strong and hard wearing. They are non-absorbent and dry more quickly than other materials. They do not shrink and do not crease readily, needing very little ironing. They are also moth-proof and resistant to acids and alkalis.

The original nylon yarn takes the temperature of the atmosphere and therefore the material woven from it is hot on hot days and cold on cool days. Nowadays synthetic filaments can be treated in various ways to overcome this drawback. There are various trade names for the different nylon materials.

Good clothing should be:

(1) *Loose and light.* Clothing must be loose and light so as to allow a layer of air to remain over the skin; not to obstruct the vital functions of the body such as circulation; and to allow the easy performance of all voluntary movement.

Shoes in particular should not be tight and should not cramp the toes. The great toe should be in line with the instep.

(2) *Warm.* Clothing should be warm so as to prevent a too great loss of body heat. The character and quality of the clothing should vary with the external temperature.

(3) *Absorbent.* The underclothes should be made of material which can absorb much moisture. If the clothing is not absorbent the air next the body becomes saturated with water and the body feels fatigued.

(4) *Non-irritating.*

(5) *Non-poisonous.* Some dyes in the clothes may have a harmful effect on the wearer, e.g. arsenic dyes.

(6) *Non-inflammable.*

General rules for the wearing of clothes

(1) The weight of the clothes should be carried on the shoulders.

(2) Clothes, especially stockings, should be changed frequently.

(3) Special night-clothes should be worn.

(4) The bedclothes should be light, but warm.

Exercise and recreation

Exercise should be taken every day, the amount varying with the person's type of work. Everyone should develop some kind of hobby; if exercise and recreation can be combined so much the better. Gardening and active games, for example, are useful both for exercise and recreation.

Exercise in the open air is better than indoor exercise, although the latter is better than none at all. Exercise has a stimulating effect on all the vital functions; it increases the tone of the muscular system, and it improves mental balance.

Rest

Rest in bed is essential for everyone. During sleep body tissues are repaired, and the digestive organs are rested.

Lack of sleep or broken sleep lead to fatigue of the body and mind, and it is a common cause of malnutrition in children, since during sleep new tissues are built up.

The amount of sleep required varies with age. On an average per 24 hours:

An infant requires 16 hours' sleep;
A child of 4–8 years requires 12 hours' sleep;
A child of 8–12 years requires 11 hours' sleep;
A child of 12–14 years requires 10 hours' sleep;
A person of 14–20 years requires 9 hours' sleep;
An adult requires 8 hours' sleep.

Invalids and old people should have as much rest as possible, since sleep is the best tonic, and in the old the repairing of tissue is slowed down.

To ensure peaceful sleep the bedroom should be well ventilated, quiet, comfortable and cheerful. The bedclothes should be light and warm, making it unnecessary to heat the room artifically. For cleanliness, washable sheets should be used; for infants, however, blankets should be placed next to the body, to prevent too rapid loss of body heat.

Social customs and habits

A chapter on Personal Hygiene would not be complete without reference to the increasing interest now being taken in investigations into the effects of social customs and habits, many of them traditional and of long standing, on the health and welfare of the community.

Examples of such habits are smoking, the consumption of alcohol and the misuse of drugs.

Smoking

The consumption of tobacco in the form of pipe tobacco, cigars and particularly cigarettes has reached enormous proportions. The offering of a cigarette is a socially acceptable practice, indicative of friendliness, and no doubt has its value.

Recently the connection between smoking – particularly cigarette smoking – and specific diseases has been the subject of much investigation.

The disease to which most attention has been devoted is lung cancer, and there is no doubt that there is a correlation between the amount of tobacco smoked, particularly cigarettes, and the incidence of this disease.

Laboratory investigation has shown that there is a carcinogenic, or cancer-producing, substance in cigarette smoke. There is also evidence of a correlation between cigarette smoking and other diseases such as heart disease, bronchitis, etc.

There can be no difference of opinion that if a person does smoke, he should do it in moderation. Health education campaigns are a means of bringing the risks of smoking to the attention of the public.

Alcohol

The consumption of beverages containing alcohol is one of the most ancient human social practices. Beverages may contain only small amounts of alcohol, e.g. light wine and beer; on the other hand, spirits – whisky, brandy, gin, etc. – may be composed of more than 50 per cent alcohol, causing a more marked effect on those who consume these drinks.

Alcoholic drinks are commonly regarded as 'stimulants', although alcohol is in fact a depressant of the higher mental faculties. The 'stimulating' effect is due to the removal of inhibition in individuals. In certain circumstances, for example when a person has to make a speech, such removal of inhibitions may have its advantages. Alcohol, however, does affect a person's judgement and the speed and accuracy of reactions to varied circumstances.

In the field of road safety the effect of alcohol on judgement is of particular importance. Investigations have shown that in a significant proportion of road accidents involving cars, the consumption of alcoholic drinks is at least one of the contributory causes. The effect of alcohol on different individuals is difficult to assess and the same amount may have quite different effects on two different individuals. In view of the gravity of the road safety problem the Road Safety Act of 1967 has defined a new offence: a driver with more than a prescribed maximum amount of alcohol in the blood must not be in charge of a car. Alcohol is absorbed into the blood stream and the amount in the blood is related to the amount of alcohol consumed. The advantage of a blood test is that it provides for objective evaluation.

The analysis of blood is a complicated procedure and for the screening test the 'breathalyser' has been introduced. The person concerned breathes into a bag containing a chemical whose colour is changed if alcohol is present above a certain proportion.

It must be realized that alcohol in excess is a poison. Excessive consumption can lead to coma, and may even have fatal results.

There is also the case of persons who develop, for various reasons, a compulsive craving for alcohol that cannot be resisted. This condition leads to physical and mental deterioration of the individual, and is one of the common causes of the breaking-up of family life.

Drug addiction

Akin to the problem of excessive drinking of alcohol is the problem of drug addiction. It has long been known that certain drugs used in medicine are habit-forming, the individual becoming addicted to a drug and unable to resist its use, with consequent physical and mental degeneration. They include the morphine group of drugs – heroin and cocaine are examples of such habit-forming drugs which are colloquially known as 'hard' drugs. Their use is regulated by the Dangerous Drugs Act which deals with such matters as prescription, safe storage, and the keeping of strict records.

In recent years the taking of drugs for non-medical purposes has become much more common. Certain recently discovered drugs have an effect on the mental processes, e.g. drugs which produce hallucination. These drugs have their place in the treatment of mental illness, but are now being taken rather for their pleasant effects. Such drugs are colloquially known as 'soft' drugs. It is not safe to assume that such drugs cannot be habit-forming; the greatest danger of their use is that they encourage the taking of the known habit-forming drugs, e.g. morphia and cocaine. Drugs other than the simplest should only be taken under medical supervision.

PRACTICAL

Examine fibres of different types of material under the microscope.

2 Diet

The diet should be adequate to supply energy for muscular and other work, and to replace broken-down protoplasm.

When planning a diet the amount of energy produced by different foodstuffs and the energy requirements of the body under varying conditions should be known. The body requires both heat energy to maintain the body temperature at 36·9°C, and mechanical energy to perform work. As all energy is ultimately transformed into heat, the energy produced by the body can be measured in terms of heat. The unit of heat is the large calorie (or Calorie), i.e. the amount of heat required to raise the temperature of 1,000 ml of water through one centigrade degree.

Measurement of the energy in food

The energy stored up in food is measured by burning a sample of known amount in an atmosphere of oxygen contained in a water-jacketed 'bomb calorimeter'. The amount of heat produced may be calculated from the rise in temperature of the surrounding water, and from this, the amount of heat liberated from 1 g of the material when it is broken down completely to carbon dioxide and water. In this way it has been found that:

1 g of carbohydrate liberates 4·2 Calories
1 g of fat   liberates 9·3 Calories
1 g of protein   liberates 5·6 Calories

In the body proteins are not completely broken down to carbon dioxide and water. Urea is excreted as the waste product of protein metabolism. As 1 g of urea on burning liberates 1·5 Calories, in the body 1 g of protein liberates 5·6–1·5, i.e. 4·1 Calories.

By analysis the composition of mixed foodstuffs may be determined and their calorific value calculated. No food is completely absorbed and small amounts of proteins, fats and carbohydrates are passed out in the faeces. In practical dietetics therefore the following heat values for the various foodstuffs are employed:

Calorific value of 1 g of carbohydrate	= 4 Calories
Calorific value of 1 g of fat	= 9 Calories
Calorific value of 1 g of protein	= 4 Calories

Measurement of the energy output of man

The energy output of man is measured directly by placing the subject in a human calorimeter. The calorimeter is a chamber with non-conducting walls and contains a pipe through which water passes. The temperature of the entering water is known, and the rise in temperature may be measured. As the speed of circulation is controlled the amount of water, and therefore the amount of heat given to the water, may be calculated. Air circulates through the chamber; the carbon dioxide and water produced are removed, and fresh oxygen is passed in.

This apparatus for measuring the energy output is clumsy and expensive, and is useless for clinical work. The energy output can, however, be obtained indirectly, by measuring the output of carbon dioxide and the intake of oxygen.

In this method the patient inspires room air which is then expired through a one way valve into a bag from which all air has been previously excluded. Samples of the air expired into the bag are analysed, and the total volume expired in a known time is measured. Since the composition of inspired air is known, the amount of carbon dioxide expired, and the amount of oxygen consumed over a given period, may be determined.

A modern apparatus has been designed which performs all the operations mechanically and records the results photographically, and thus any change in oxygen intake and carbon dioxide output under different circumstances may be followed.

Basal metabolism

Basal metabolism is the amount of energy necessary to keep the body alive, that is, to maintain the functions of the body when it is lying still and warm and without food.

When food is not being taken in energy is obtained by the oxidation of food stored in the body. The intake of food stimulates the metabolism of

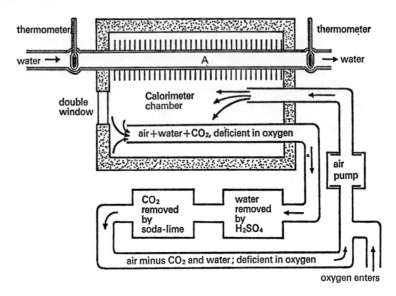

Fig. 131. Atwater Benedict human calorimeter.

the body; for this reason the basal metabolism should be measured twelve hours after the last meal. During starvation and under-nutrition the basal metabolic rate becomes very much reduced.

The basal metabolism varies with the body surface. On the whole short fat people have a lower basal metabolism than tall thin ones, who expose relatively more body surface and thus lose heat more rapidly. Men have a higher basal metabolism than women, and the basal metabolism is higher in the young than in the adult. For this reason, and also to supply material for growth, young children require relatively more food than adults.

The average basal metabolic rate per square metre of body surface per hour in Calories is given in the following table:

Age	Male	Female
13–14	46	43
20–29	39·5	37
40–49	38	36
50–59	37	35

The surface area of a person is a factor of the height and weight, and may be obtained from a graph. The basal energy output of the healthy individual may be calculated from the above table, and the amount of food required to keep the body alive may be determined. These figures vary widely in disease, so that direct estimation must be made in such a case.

The normal individual requires additional energy for carrying out the daily work. The amount of this energy necessary varies with the occupation, a manual worker requiring more than a sedentary worker.

By making a subject do graded work in a human calorimeter the energy requirements for each type of work may be determined. The energy requirements for a normal day are as follows:

8 hours' sleep (at B.M. 71 per hour)	568 Cal
8 hours awake (at B.M. 71 +30% for moving about, eating, etc.)	736 Cal
8 hours' work (at B.M. 71 +1,000 C. for light external work)	1,568 Cal
Total	2,872 Cal

As some additional work may be done, for example, in travelling to work, 3,000 Calories is generally considered to be the energy output of an average working man per day. A sedentary worker requires less energy and a heavy manual worker more. The following table shows the approximate energy requirements of different types of workers:

	Calories
Man doing light work	3,000
Man doing medium work	3,500
Man doing hard work	4,500–5,500
Woman doing light work	2,500
Children 0–6 (both sexes)	1,500
6–10 (both sexes)	1,800
10–14 (both sexes)	2,500
Girls 14–21	2,500
Boys 14–21	3,000

The foodstuffs must not only supply sufficient energy, but carbohydrates, proteins and fats should be present in the diet in the correct proportions.

Proteins

Proteins are essential for diet. On digestion they produce amino acids, which build up new protoplasm. Of the amino acids used by the body eight have been proved to be essential to life; omission of any one in the proteins of the diet causes lack of growth and loss in weight. It is necessary to eat some of the proteins which supply on digestion the essential amino acids.

Proteins may be divided into:

First-class proteins found in meat, fish, egg, milk and cheese. These contain all the essential amino acids.

Second-class proteins of vegetable origin. These are not so valuable – some, e.g. zein of maize and gelatin, are poor in the essential amino acids.

During the early stages of starvation 7–12 g of nitrogen, corresponding to 44–75 g of body protein, are excreted per day. This excretion also occurs when a person is taking a normal diet, and if the diet is to make good this loss of nitrogen it must contain 75 g of protein. To ensure that the protein contains enough of all the essential amino acids 100 g of proteins are required. As vegetable proteins are second-class proteins, vegetarians require more protein.

As proteins may be used as a source of energy, it would be possible to exist on a pure protein diet. A great deal of protein would have to be eaten and such a diet would be unappetizing, indigestible and expensive.

For example, a heavy manual worker requires 4,500 Calories per day. If this were supplied by meat alone he would require 1,000 g of protein, i.e. 5,000 g or 11 lb of meat per day.

If, however, a man eats 100 g of protein per day as suggested it supplies him with 400 Calories. The remaining Calories could be supplied by carbohydrates, fats or a mixture of both.

Carbohydrates and fats

Although fat produces much energy, it cannot be completely burnt in absence of carbohydrates. Two grammes of fat require one gramme of carbohydrate for satisfactory oxidation. If carbohydrate is not available poisonous substances known as ketones are produced. Carbohydrates are said to 'light the fires on which the fat burns'.

If the fat in the diet falls below 50 g per day there is a definite craving for it. The amount of fat consumed depends on the person's 'appetite for fat'. It is a less bulky food than carbohydrates, containing less water, and weight for weight it produces twice as much energy. It has more 'staying' power, as it takes longer to digest. Half a slice of bread is digested in $1\frac{1}{2}$ hours, while a piece of bread and butter is only digested after 4 hours. Fat supplies energy more slowly than carbohydrates.

The whole of the energy requirements may be obtained from carbohydrates but such a diet would be indigestible on account of its bulk.

When the diet is in excess of immediate requirements, carbohydrates may be converted into fat and stored.

In a normal daily diet an average of 100–125 g of fat are taken, the remaining calories being made good by eating carbohydrates.

The following is an example of a well balanced diet:

	g	Calories
Protein	125	500
Fat	125	1,125
Carbohydrate	400	1,600
		3,225

When planning a diet it is necessary to know the composition of various foodstuffs. The composition of some of the common ones is given on page 264.

Mineral salts

Mineral salts are an essential constituent of a complete diet. That common salt or sodium chloride is necessary is well known. There are in addition numerous other important mineral constituents. These include mineral salts containing calcium, phosphorus, magnesium, potassium, iron, chlorine and iodine.

Calcium is important in the formation of teeth and bone; about 27 per cent of bone is calcium. The majority of people do not take in enough calcium in the diet, about 0·75–1 g being required per day. More is required in pregnancy.

Lack of calcium in the diet is a cause of rickets. Calcium is present in most tissue fluids, green vegetables and milk. Butter is poor in calcium, as this element is not present in the cream from which butter is made. Skimmed milk is rich in calcium. Bread and meat contain only a small amount of calcium.

In teeth and bone, **phosphorus** is found combined with calcium in salts known as calcium phosphates. It is taken in the diet in certain proteins, e.g. casein, nucleo-proteins.

Magnesium is necessary for the formation of the skeleton. Rats fed on a diet deficient in magnesium suffer from a form of rickets. Magnesium is also essential for normal cell functioning. Meat is the chief source of magnesium in the diet.

Sodium chloride and potassium chloride. Sodium is taken in, in combination with chlorine, in the form of common salt or sodium chloride. Sodium chloride is an important constituent of blood, keeping the osmotic pressure of the blood and tissue fluids constant, and the cells are bathed in an unchanging medium.

A considerable amount of salt is excreted by the kidneys and sweat glands (see page 147), and in hot countries where a great deal of salt is lost in the sweat and large quantities of water are drunk, the diet must contain more salt.

Sodium chloride is necessary to provide the gastric glands with material for building up their hydrochloric acid. Lack of sodium chloride in the diet produces muscular cramp and may produce indigestion.

Potassium is taken in vegetable foods, in combination with chlorine. It helps to maintain the intracellular osmotic pressure.

In the body there is a balance between the sodium and the potassium. If no sodium is taken, the sodium chloride in the blood is replaced by potassium chloride. This has a harmful effect on the heart.

Iron must be taken in, in the food, as it is an important constituent of haemoglobin, the colouring matter of blood. Lack of iron in the diet leads to an anaemia known as nutritional anaemia. This can be cured by taking in a large dose of an iron compound. Milk is lacking in iron. Babies are born with a store of iron in their liver. If, however, this store is deficient they suffer from anaemia. Anaemia is common in adolescent girls and pregnant women.

Iron is present in meat (especially in liver), in green vegetables and many fruits.

Iodine is necessary for the formation of thyroxin, the secretion of the thyroid glands (see page 205). Sea fish is rich in this element.

Trace elements

A number of other elements such as fluorine, manganese, cobalt, zinc, copper, are required in minute quantities for special purposes. Fluorine, for example, is essential for the proper formation of the teeth. These 'trace elements', as they are called, are usually present in a diet which is adequate in other respects.

Vitamins

Until the end of the nineteenth century it was thought that a perfect diet contained carbohydrates, fats and proteins, mineral salts, water and roughage, but when animals were fed on these substances in an absolutely pure state they did not thrive. This was thought to be due to the unappetizing quality of the mixture.

Advances in the investigations of a complete diet are due to observations which at the time appeared to have no importance.

Up to the sixteenth and seventeenth centuries scurvy was very rife on board ship, and it was found that this disease would be cured by taking the juice of citrus fruits. In 1865 a Board of Trade regulation was brought into force which required all ships to carry a supply of lemons and limes to add to their drinking water.

In 1887 the Dutch Government established a laboratory in Java to study beri-beri, a disease of rice eaters which produces degeneration of nerve endings and symptoms of paralysis. It was accidentally discovered that hens and pigeons fed on polished rice developed similar symptoms. When they were fed on unpolished rice or polished rice with the husks added the birds recovered. It was discovered that the anti-beri-beri substance could be extracted from the husks with water or alcohol. It was thought that some substance in the husks acted as an antidote to food poisons present in the rice.

During the first decade of the twentieth century Sir Frederick Gowland Hopkins experimented on rats, feeding them upon a mixture of pure protein, fat and carbohydrate with the necessary mineral salts added. The rats did not grow and became sickly. When a small amount of milk was added to the purified diet the rats recovered and developed normally. Hopkins published a paper in 1912 in which he stated that there must be certain accessory factors in a diet which are only present in small quantities. Cashmir Funk continued the work on beri-beri, and tried to isolate the anti-beri-beri substance. He obtained a crystalline substance which had curative powers on birds suffering from polyneuritis. This he named vita-amine, since it was necessary to normal life and contained an amine group.

It was obvious that absence of some factors were concerned with scurvy, beri-beri and rickets (it had been known for some time that cod-liver oil had a definite curative effect on rickets), and these food factors were all called vita-amines. As it was proved that some at least contained no nitrogen the name was changed to vitamin. Since then at least seven vitamins have been discovered. They are called:

Vitamin A (antixerophthalmic vitamin)
Vitamin B complex
Vitamin C (ascorbic acid, or antiscorbutic vitamin)
Vitamin D (antirachitic vitamin)
Vitamin E (antisterility vitamin)
Vitamin K (antihaemorrhage vitamin)

Vitamin A

This is a fat-soluble vitamin found in milk, fat, butter, liver oils and egg yolk. Cod-liver oil and halibut oil are especially rich in this vitamin. Vitamin A is not found in fat stored in plants, and for this reason vegetable margarine cannot be a complete substitute for butter, unless vitamins are added.

Vitamin A can be manufactured in the body from the pro-vitamin carotene, a pigment associated with chlorophyll in green leaves. It is destroyed slowly by heat, and there is some loss when milk is heated and vegetables cooked.

When vitamin A is lacking in a diet there is a drying up of the mucous membranes. This affects in particular the eyes, so that a disease known as xerophthalmia, in which the lids and cornea become keratinized, is produced. The drying up of mucous membranes makes the body less resistant to disease, especially those which affect the bronchial tracts and alimentary canal.

Without vitamin A visual purple in the retina cannot be regenerated and night blindness results. A great number of people suffered from this during and just after World War I, when their diet was poor in fats and fresh vegetables. It is still common in some countries, such as China, Japan and India.

Children lacking vitamin A in their diet do not grow properly. This lack of vitamin may be made good by giving cod-liver oil. Vitamin A is often spoken of as the **'growth vitamin'**.

Vitamin D

Vitamin D is also a fat-soluble vitamin. It is found in cod-liver oil, egg yolk, butter and cream. Lack of it in the diet of children leads to a disease called rickets, in which the bones of the body become deformed. The teeth are very soft and decay easily. This disease is common among the poorer people in temperate climates, and its incidence is more marked in winter than in summer.

Lack of vitamin D in the diet of adults leads to a disease called osteomalacia or softening of the bones.

Without vitamin D the calcium and phosphorus cannot be laid down in the form of calcium phosphates in the bones and teeth. The structural part of the body is in a constant state of flux, and is not permanent from the time it is formed in childhood. It has been shown by X-ray analysis that 20 per cent of the calcium of bones can be replaced within 30 days.

It has been known for a long time that children who spend some of their time out of doors playing in the sun, however poor their diet, never suffer from rickets. This led investigators to realize that light must have some effect on the disease. It is now known that vitamin D can be manufactured in the body from a pro-vitamin ergosterol, by the action of the ultra-violet light of the sun. Nowadays ultra-violet light is used in the treatment of children suffering from this disease.

If an excessive amount of vitamin D is taken in the diet the salts of bone are laid down in other parts of the body, e.g. the arteries; this is dangerous. A moderate overdose is not harmful.

Vitamin E

Vitamin E is the third fat-soluble vitamin. It is found in cereals, meat, lettuces, and is not destroyed by cooking.

The effect of its absence from the diet of human beings is still uncertain, but in rats it leads to sterility in the male, and re-absorption of the foetus in the female.

Vitamin B complex

This is a large group of water-soluble vitamins which are found together and for this reason for a long time escaped identification as separate substances. The B complex is found in yeast, liver, milk, green vegetables and whole-meal flour. Purified carbohydrate food, e.g. white flour and sugar, is deficient in it.

Three vitamins of the B complex have been shown to be indispensable in the human diet; they are vitamin B_1 (antineuritic vitamin), vitamin B_2 (riboflavin) and nicotinic acid. Lack of B_1 in the diet leads to the diseases already described called beri-beri, or in birds avian polyneuritis. Lack of nicotinic acid leads to a skin disease, pellagra, of which the symptoms are inflammation of the skin and nervous disorders. Ribo-flavin is thought to play a part along with vitamin B_1 in carbohydrate metabolism.

Vitamin C

Vitamin C is another water-soluble vitamin, found in fresh fruits and lightly cooked vegetables. Boiling for any length of time destroys the vitamin. As the vitamin comes out in water, green vegetables should be cooked in a small amount of water. They should also be brought to the boil quickly in order to destroy the enzyme which breaks down the vitamin. Vitamin C has been isolated and found to be identical with a substance called ascorbic acid.

Lack of this vitamin in the diet leads to scurvy, symptoms of which are spongy gums, and bleeding in all parts of the body, especially the muscles, joints and internal organs. A slight lack leads to a condition of sub-scurvy, characterized by an 'off-colour' feeling, irritability, laziness and argumentativeness. Explorers suffered from scurvy in the middle ages and practically all peasant communities living on salt meat during the winter suffered from sub-scurvy towards the end of the winter. There were outbreaks in belligerent countries during the First World War, and the outbreaks in England in 1917 were probably due to lack of potatoes.

Babies fed on artificial and sterilized diet get infantile scurvy.

Nowadays ships take supplies of fruit juices and explorers can obtain ascorbic acid in tabloid form.

Vitamin K

Vitamin K is fat-soluble and seldom lacking from the diet. It is necessary for the clotting of blood.

Besides carbohydrates, fats, proteins, mineral salts and vitamins, for a complete diet a person must take in water and roughage.

Water is an essential constituent of protoplasm, and is necessary for the body fluids. It is continually being lost from the body in the sweat, urine, and from the lungs, and must be replaced.

Roughage has been mentioned on page 111. It is the indigestible part or cellulose of the food, and it forms bulk in the alimentary canal. This stimulates the muscles in the wall of the alimentary canal and promotes peristalsis. If food remains for any length of time in the canal toxins are produced by the action of the bacteria always present, and these poisons pass into the blood.

Cooking of food

Food is cooked to make it more palatable, more digestible, and to kill any germs or eggs of parasites which may be present in the uncooked food.

A disadvantage of cooking is that it destroys most of vitamin C present in the raw food. Vitamins A, B, D and E are fairly stable and only destroyed to a small extent. If soda is added, as, for example, in the cooking of vegetables, the vitamins will be destroyed.

Cooked food becomes more digestible, certain changes being brought about by the heat.

When meat is cooked the connective tissues between the muscle fibres becomes softened and the meat becomes tender. Prolonged cooking coagulates the proteins and the meat becomes hard and indigestible.

In vegetables the heat bursts the indigestible cellulose cell walls and sets free the sap or starch grains which then burst within the cells.

Cooking may be carried out in seven ways: boiling, stewing, frying, grilling, roasting, baking and steaming. For a description of these methods the reader is referred to an elementary cookery book.

Pressure cooking

In a pressure cooker a higher temperature is obtained than in an ordinary container and therefore food cooks more quickly. This in itself may be an advantage and during pressure cooking there may be less loss of nourishment than in other methods since only a small amount of water is used and a high temperature is less destructive than a long cooking time.

The essentials of a complete diet can now be summarized:

(1) It must contain enough food to supply the necessary calories.

(2) It must contain the three classes of foodstuffs, namely carbohydrates, fats and proteins, in a correct proportion. Some animal protein should be included to be sure that the body is supplied with the essential amino acids.

(3) It must contain plenty of fruit and green vegetables (some uncooked), to supply the body with roughage, vitamins and mineral salts.

(4) It must be palatable. Food is rendered more palatable by good cooking, good serving and sometimes by the use of condiments. When food is palatable it stimulates the proper flow of the digestive juices.

There is some truth in the saying, 'A little of what you fancy does you good.'

3 Some important foodstuffs and food preservation

The average composition of some of the commoner foodstuffs is given in the following table:

Food	% Water	% Protein	% Carbo-hydrate	% Fat	% Mineral Salts	Calories per lb
Bread (white)	39	10·6	48·5	1	0·9	1,117
Rice (boiled)	72·5	3·0	24·5	—	—	498
Rice (uncooked)	12·3	8·0	79·7	—	—	1,591
Carrots	88·5	1·0	9·0	0·5	1·0	205
Lettuce	94·5	1·5	3·0	—	1·0	87
Potatoes (skinned and boiled)	75·5	2·5	21	—	1·0	429
Potatoes (raw)	78·3	2·5	18·5	—	0·7	378
Egg	73·5	13·5	—	10·5	1·0	74
Cheese	35	25	—	34·5	5·5	2,150
Mutton (leg, lean)	67·5	20	—	11·5	1·0	1,086

Cheese

Cheese is made from sour milk by pressing out the whey, and allowing bacteria to grow in the remaining solid mass. The variety of cheese formed depends on the species of bacteria added. Cheese has a very high dietetic value. It is, however, rather indigestible, as the fat forms a covering round the protein casein, preventing the digestive juices from acting on it until the fat is broken up.

Meat

Meat is chiefly the muscular tissue of animals, and therefore contains a high percentage of protein and fat. The vitamin content is low. Beef and mutton contain about 20 per cent protein and 65–70 per cent water. The amount of fat varies with the age of the animal, its nutrition, and the 'cut' of the joint. The proteins are Class I (see page 256).

The muscular tissue of poultry is white and contains less fat and less connective tissue. White meat is more easily digested than red meat.

Meat is also obtained from other organs of animals, e.g. liver, kidneys, pancreas (sweet bread), brain. This meat has proteins which are superior to the muscle proteins, and has a higher vitamin content. Liver contains much iron.

Fish

Fish has a similar composition to other kinds of muscular tissue. It contains about 10 per cent of protein of equal biological value to meat protein, and a varying amount of fat. A herring, for example, contains about 20 per cent fat and therefore has a high calorific value. The whiting contains only 2 per cent of fat, and has a low calorific value, but is more easily digested.

Fish contains iodine, required by the body for the formation of thyroxine; it also contains vitamin D and other fat-soluble vitamins. Fish oils, e.g. cod-liver oil, are used to make good any vitamin deficiency in the diet.

Eggs

The average composition of an egg is protein 14 per cent, fat 11 per cent and water 74 per cent. The white consists of a fluid protein contained in a network of connective tissue; the yolk consists of a protein containing phosphorus and sulphur, together with the fat.

Since eggs have a high vitamin content, a high protein content, and are very digestible, they form an important food. The more an egg is cooked, the less digestible its proteins become.

Flour

Flour is used in baking bread, pastries, cakes, etc. It is made by grinding up wheat-grains. If part of the husk is left on the grain, wholemeal flour is formed; if the husks are completely removed brown flour is formed; if the brown portions of the grains within the husks are removed white flour is formed.

All flours contain a high percentage of starch and a certain amount of protein. Brown flour contains more protein and vitamins than white flour, since the protein and the vitamins lie under the brown outer covering; wholemeal flour in addition contains roughage.

Flour protein contains gliadin, a protein which becomes sticky when wet, and enables flour to be made into a dough.

Bread

Bread is made from flour dough by adding yeast and salt. The yeast converts some of the starch into sugar which it ferments. The carbon dioxide produced causes the bread to rise. The dough is then cooked in the oven. This baking stops the fermentation.

Milk

Milk is a food containing all the essential constituents of a complete diet, the fat being present in an emulsified form. It is the correct diet for infants, but for adults it is too watery to be satisfying, and contains no roughage.

Human milk and cow's milk are not identical in composition. They are compared in the following table of percentages:

	Cow's milk	Human milk
Water	87·3	88·4
Sugar (lactose)	4·5	6·5
Protein (caseinogen lactalbumen)	3·5	1·5
Fat	4	3·3
Salts (Phosphates, Ca, K, Mg)	0·7	0·3
	100%	100%

Cow's milk contains less sugar, but more protein than human milk. It contains more salts than human milk.

When the digestive juices of an infant act on the protein of human milk, a loose, easily digested curd is formed. When they act on cow's milk the curd formed is harder and less easily digested.

The fat in human milk is in a much finer state of division than the fat of cow's milk, and it is therefore more easily digested.

Cow's milk may be given to infants as a substitute for human milk if it is diluted and lactose added. Sometimes sodium citrate is added to make the milk protein more digestible. The milk should be boiled for infants and young children.

Cow's milk

The amount and quality of the milk varies from cow to cow, and it also varies throughout the year. Generally the fat and salt content of milk is lower in the summer than in the winter. An unhealthy cow gives milk low in protein and fat.

Under the Food and Drugs Act (1938) milkmen may be prosecuted for selling milk containing less than 3 per cent of fat and less than 8·5 per cent of non-fatty solids. The common form of adulteration of milk is removal of cream and addition of water.

All milk should be produced under clean conditions. There are two types of foreign matter which may enter the milk:

(1) Dirt from the byre, cows, milkers, utensils, etc.

(2) Micro-organisms. Some organisms, e.g. the non-pathogenic *Bacillus lactis,* are normally present. *Bacillus lactis* converts the lactose into lactic acid and sours milk. When a certain degree of acidity is reached the caseinogen curdles. This souring is delayed if the milk is rapidly cooled after milking.

Milk forms a good medium for growth of most micro-organisms, and infected milk is responsible for many outbreaks of disease.

Production of clean milk

Milk is liable to contamination either at the time of production or during its distribution. The Ministry of Agriculture and Fisheries and local health authorities have been given powers under the Food and Drugs legislation to enforce a number of requirements at both of these stages. The main ones are:

(1) Cows must be healthy and if diseased must be removed from the herd and their milk must not be sold. They must be kept clean and well groomed, and their flanks and udders must be washed before milking.

(2) Milkers must keep themselves and their clothing in a clean condition. They may be required to cease work if they have been in contact with infectious disease. Milking machines are now used in modern dairy farms.

(3) Cow-sheds used for milking must be ventilated and have enough light to enable milking to be carried on. The floors should be sloped and constructed of some impervious material, facilitating drainage and

cleansing. Water should be available. No pigs or poultry should be allowed in cow-sheds where milk is produced. Manure, etc., must be removed from the vicinity of the shed.

(4) Churns, etc., must be so constructed as to ensure thorough cleansing. They must be scalded with boiling water or steam as soon as possible after use. Much milk is now bottled at the farm.

(5) No room in which milk is handled should communicate directly with a W.C. or have a drain opening into it. A living-room should not be used for the handling of milk.

(6) All milk producers must register with the Local Health Authority, to facilitate their control.

The Milk and Dairies Order applies to all milk producers and its aim is to ensure that milk is produced in conditions in which the risk of contamination is reduced as far as possible. The order does not lay down any hygienic standard for milk itself after it has been produced, and it does not require cattle to be examined at regular intervals by veterinary surgeons – the only method of eliminating unhealthy – especially tuberculous – cattle.

Tuberculosis is a disease which is frequently spread by milk. Cows suffer from the disease, and if the udder is infected the organism of tuberculosis may be secreted in the milk. Such milk appears of good quality, but is liable to cause the disease in human beings.

Quite apart from cleanliness in the production of milk, it is essential that it should be free from disease-producing organisms. To encourage the production of clean milk and of milk free from tuberculosis infection, a graded milk scheme has been introduced. A producer whose milk reaches certain hygienic standards and whose herd is composed of healthy animals receives a higher price for his milk than the ordinary producer.

The different designations or grades of milk, and the appropriate requirements for each, are set out in regulations issued in the Food and Drugs (Milk, Dairy and Artificial Creams) Act 1949, an Act which consolidates the main enactments relating to milk and its products. These requirements for designated milks are, of course, in addition to those designed to ensure cleanliness in production. The principal grades of milk are:

(1) **Tuberculin tested milk.** This milk is produced from herds kept isolated from other herds. The animals of the isolated herds have been examined by a veterinary surgeon and have passed at stated intervals a

'tuberculin test'. This test consists of injecting dead organisms of tuberculosis into the animal. If a rise in temperature follows, the reaction indicates infection of the animal with tuberculosis and such an animal must be removed from the herd.

Such milk must be labelled as 'Tuberculin Tested'. If bottled on the farm where it is produced it may be labelled in addition 'Farm Bottled'. In Scotland such milk is labelled 'Certified'.

(2) **Accredited milk.** This milk is produced from cows in a herd which is kept separate from other animals and is subject to veterinary examination. No diseased animals may be retained in the herd. The herd is not subject to the tuberculin test but any known reactor must be excluded.

Both these milks have to satisfy bacteriological tests which are designed to give an indication of the number of organisms present, and thus maintain the standards of cleanliness in production. The present method is to estimate the number of such organisms indirectly by their effect on an aniline dye, methylene blue; this test is more rapidly and easily performed than the older one of estimating the actual number of organisms.

'Tuberculin Tested' milk may be considered as free from infection with tuberculosis. 'Accredited Milk', while produced in conditions of cleanliness and from animals clinically healthy, is not so 'safe' as tuberculin tested milk. It is possible for a cow showing no clinical signs to be infected with tuberculosis.

Both these milks are known as 'raw' milk and are sold to the public untreated by heat. Heating milk destroys disease-producing organisms and in this way milk may be made 'safe'. Such heat treatment has, however, to be efficiently carried out and the Milk (Special Designations, Pasteurized and Sterilized Milk) Order lays down the necessary conditions and appropriate laboratory controls.

Heat-treated milk may be:

(*a*) *Pasteurized.* In this process the milk is raised to a temperature of 71°C for 15 seconds and then rapidly cooled to not more than 10°C. Recording instruments to ensure compliance with these conditions are essential.

(*b*) *Sterilized.* The milk is heated to boiling-point, bottled, distributed and sold in hermetically sealed bottles.

The provisions of the Milk and Dairies Order are enforced by Local Health Authorities; these authorities also supervise pasteurizing plants.

The provisions of the Special Designations Order are enforced by County Councils and County Borough Councils.

The Milk and Dairies Order lays down conditions under which all milk must be produced. The Special Designations Order lays down a standard which graded milk must meet.

For infants and young children all milk, including tuberculin tested and pasteurized, should be boiled immediately before use and cooled quickly. Boiling kills the germs which may have entered the milk from handling and storing after the bottle has been opened.

Preservation of food

Food 'goes bad' through the action of bacteria, which decompose it. The basis of food preservation is to produce conditions unfavourable to the growth of these organisms.

Bacteria for multiplication must have:
(1) A suitable temperature.
(2) Moisture.
(3) A certain amount of air.
(4) Absence of chemicals which poison them.

Methods of food preservation adopted are:

(1) **Exposure to low temperatures and freezing.** Food may be kept for a long time at low temperatures. This method is used on a small scale in hospitals, large institutions and private houses, as refrigerators are now comparatively cheap.

In recent years, the 'deep freeze' apparatus which keeps food at a temperature well below its freezing point has come into use in food storage plants, food shops, etc., including private houses, and enables food to be preserved for long periods.

Meat may be chilled or frozen. Chilled meat will keep for five or six weeks, while frozen meat will keep for a much longer time. Frozen meat is brought to this country from Australia and New Zealand, and arrives in a good condition. It is darker in colour than fresh meat and it should be cooked as soon as it has thawed.

Exposure to low temperatures does not kill the bacteria, but it prevents their activity.

(2) **Chemicals.** Certain food substances are preserved by salting, e.g. fish, ham; common salt or saltpetre being used.

In jam-making the fruit is boiled and then sugar is added to preserve it.

Two chemical preservatives sometimes used are sulphur dioxide and benzoic acid. Benzoic acid, for example, may be added to pickles and sauces; sulphur dioxide may be added to sausages. In both these cases the presence and quantities of these substances used must be indicated on the label.

It is now illegal to use any other preservatives, such as boric acid (which used to be added to cream), or salicylic acid.

(3) **Drying.** Certain eatables, e.g. fruits, milk, etc., can be preserved by removing the greater part of their moisture. This prevents the activity of the bacteria, though it may also destroy some of the vitamins, especially vitamin C.

Dried milk is an important substitute for cow's milk. It is clean and digestible.

(4) **Smoking.** After some food substances, e.g. ham, have been salted, they are smoked. This forms a dry layer on the outside, and the smoke which penetrates the outer layers has also an antiseptic effect.

(5) **Canning.** The food to be canned is placed in a tin which is closed except for a small hole. The tin and contents are then heated to remove the air and then the hole is closed by solder to prevent air entering. The tin is then sterilized by heating to a high temperature. If the food has been properly canned, this is the safest form of stored food and does not affect the vitamin content to any great extent. Once a tin has been opened the contents should be transferred to some other kind of vessel.

If the sterilization or the soldering is imperfectly carried out, the contents of the tin decompose and produce gases. These cause the tin to bulge producing a 'blown' tin, the contents of which should not be eaten. Tinned food should only be eaten if it appears to be in good condition when the tin is opened.

Food poisoning

The term 'food poisoning' is given to the group of diseases which arise from the eating of food containing harmful substances. Attacks of food poisoning are characterized by nausea, abdominal pain, diarrhoea and vomiting, and if the attack is severe, by headache, giddiness and collapse.

The patient's symptoms may vary from mild indigestion to severe abdominal pain or even fatal collapse.

The poisonous substances may be organic or inorganic.

Inorganic

The harmful substances gain entrance to food during its preparation or distribution. Tinned foods are liable to be contaminated with lead from the solder used during the canning process; fruit juices may attack the enamel of cheap pans. An outbreak of arsenic poisoning was caused by the use of a substance, containing arsenic as an impurity, in the brewing of beer.

Organic

(1) Persons may be poisoned by eating poisonous plants or berries; such accidents are most common amongst children. Poisonous fungi are frequently eaten in mistake for mushrooms.

(2) Poisoning may be due to the presence of bacteria, or toxins produced by bacteria, in the food. Food may be contaminated at any stage of its preparation by the bacteria which cause disease. The most important group of organisms causing food poisoning are known as the 'Salmonella' group. Many are normal inhabitants of the intestine of animals; some cause diseases very similar to enteric fever.

While cooking may destroy the germs of food poisoning it may leave the toxin unaffected and capable of causing the symptoms of the disease. Cases due to toxins have a more rapid onset than cases due to infection with bacteria; they usually recover rapidly.

Tinned foods and prepared meats such as brawn, sausage, tinned salmon, are the foods most commonly involved in food-poisoning outbreaks. A number of outbreaks have been traced to duck eggs, either eaten raw or lightly cooked; such eggs ought always to be thoroughly cooked before being eaten.

Large quantities of tinned foods are imported from abroad and consignments are subject to inspection at the port of entry in order to detect tins which are in any way defective, leaking, etc., and any tins that show signs of ineffective canning, e.g. tins being blown, which indicates some form of decomposition of the contents.

High standards of hygiene must be maintained throughout the canning process. One major outbreak of typhoid in the United Kingdom was traced to the use of contaminated water in the cooling of tins during the canning process.

Botulism is a special type of food poisoning which is fortunately rare in this country. It is due to a toxin produced by a bacterium of the species *Clostridium botulinum*. The toxin produces respiratory paralysis and the disease is usually fatal. The well-known outbreak at Loch Maree in Scotland in 1922 was traced to potted meat.

PRACTICAL

1 Experiments on milk.
   (*a*) Examine a drop of milk under the microscope. Note the fat globules.
   (*b*) Allow some milk to stand. The cream or milk rises to the surface.
   (*c*) Leave some milk to sour. Curds are formed. Test these for protein, and test the clear liquid or whey which separates for a reducing sugar.
   (*d*) Add some rennin to a small quantity of milk.
2 Test cheese, meat, fish, flour, bread, carrot, potato, etc., for starch, sugars, fat and protein (see Part I, Chapter 11).

4 Air and ventilation

The main constituents of atmospheric air are, approximately, by volume:

Nitrogen	78 per cent
Oxygen	21 per cent
Carbon dioxide	0·03 per cent

It also contains about 1 per cent of moisture. The carbon dioxide content varies greatly and may be as much as 0·4 per cent in ill-ventilated rooms.

The contents of the air of occupied rooms is influenced by the products of respiration. Expired air contains about 4 per cent of carbon dioxide, and the oxygen content is decreased; it also is saturated with moisture. Micro-organisms and organic excretions are expelled during respiration. The temperature of expired air is equal to body temperature.

Atmospheric pollution in large towns is due to the products of combustion from both domestic fires and industrial processes. The waste gases are carbon dioxide, carbon monoxide, sulphur dioxide, hydrogen sulphide, ammonia, etc.

Smoke reduces the amount of ultra-violet light, increases fog, and causes damage to buildings and vegetation. Smoke may be diminished by the abolition of open fires, the complete combustion of fuel or the use of smokeless fuel.

Ventilation

Ventilation is the science of maintaining atmospheric conditions which are comfortable to the human body. At one time it was thought that the discomfort of badly ventilated rooms was due to lack of oxygen, and to excess of carbon dioxide. Recent work has, however, shown that this is not so, otherwise human beings could not live at a height of over 5,000 ft, where the oxygen content of the air is much below that of the worst ventilated room, nor men work in submarines where the carbon dioxide content of the air may reach 3–5 per cent – about ten times that of ill-ventilated rooms.

It is now known that the comfort of an atmosphere depends on three factors:

(1) Temperature.
(2) Humidity.
(3) Movement of air.

When a room is badly ventilated the temperature of the atmosphere is raised. The air stagnates and the air immediately surrounding the body – that enmeshed in clothing – becomes saturated with moisture. There are thus developed the three conditions that lead to discomfort – heat, humidity, and stagnation of air.

The conditions required for comfort are:

(1) A temperature of about 15·5°C.
(2) A relative humidity not more than 75.
(3) A current of air of about 2–3 ft per second.

The aim of ventilation is to supply fresh cool non-saturated air to a room without causing discomfort from draughts. It is said that the air of a room may change six times in an hour without a feeling of draught being produced.

Various standards have been laid down as to the number of persons who can be accommodated in rooms. The standards may be stated as the number of cubic feet per person, but a more useful yardarm is the floor space per person.

Standards vary very much. Infectious diseases hospitals allow 144 ft^2 per bed. The Merchant Shipping Act insists on 120 ft^3 per seaman.

Methods of ventilation

Ventilation may be natural or mechanical.

(1) **Natural ventilation** is dependent on natural forces.

(a) *Diffusion* – the natural tendency for gases to intermix with each other.

(b) *Winds* – which may act by blowing directly through rooms, etc., or by aspiration, a suction caused by wind blowing across the top of the chimney and ventilating flues. The disadvantage of winds in ventilation is their irregular action.

(c) *The movement of masses of air of unequal temperature.* Hot air rises and is replaced by cold, relatively heavy air. These convection currents are the chief factor at work in natural ventilation. The greater the difference in temperature the greater will be the movement of air.

Dwelling-houses are ventilated by means of doors, windows and chimneys, and the floor space per person in the normal dwelling-house is usually sufficient to allow an adequate supply of air by this means. The open fire is one of the greatest aids to ventilation as it starts convection currents, the heated air passing up the chimney, and fresh air being drawn into the room to replace it. Housing by-laws require every habitable room to have one or more windows, and to be provided with a flue or airshaft with a minimum section of 100 in².

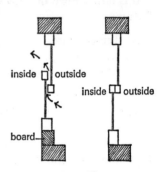

Fig. 132.
Hincke's bird device.

Windows may be casement windows opening on hinges or double-sash windows moving on pulleys. In sash windows not more than half the window area can be open at one time. If a board is fitted to the bottom of the lower sash, air can enter where the sashes meet, and the window can never be completely sealed.

Various devices have been introduced with the aim of ensuring ventilation without producing draughts from open windows. They may be in the form of:

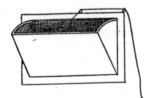

Fig. 133. Sheringham's valve.

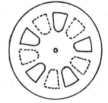

Fig. 134. Cooper's disc.

(a) *Hoppers* in one or more window-frame cheeks to prevent draughts.

(b) *Panes* with regular openings, arranged within a circle which may be closed by a glass or metal disc (Cooper's disc).

In public halls, schools and factories, windows and chimneys may not be sufficient to ensure satisfactory ventilation and additional outlets and inlets for air are required. These include:

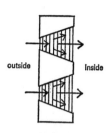

(*a*) *Perforated bricks* built into walls, e.g. Ellison's bricks. These are usually placed in outer walls just above the damp-proof course and are used to ventilate spaces under floors. These are also found in walls of private houses.

(*b*) *Wall openings*, e.g. the Sheringham valve and the Tobin's tube. Although more complicated they have no advantage over the simple perforated brick.

Fig. 135. Section through Ellison's bricks.

(*c*) *Roof ventilators*. These are effective in halls, etc. They are most useful for the extraction of air, as the hot air rises.

An example of such an apparatus is McKinnel's ventilator in which a source of heat at the base of the central shaft causes an upward current of air. In windy weather air may enter by them, causing down draughts.

Natural ventilation can be greatly assisted by the judicious location of radiators, etc. If they are placed near an air inlet such as a perforated brick they warm the incoming air and assist the passage of air through the ventilator.

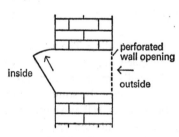

Fig. 136. Side view of Sheringham's valve.

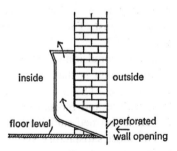

Fig. 137. Side view of Tobin's tube.

Air stagnation is one of the causes of discomfort in closed and crowded rooms, and electric fans, by keeping the air in motion, may give much comfort to the occupants. In order to allow an adequate circulation of air round houses, by-laws regulate the number of houses which can be built to the acre and lay down the minimum breadth for roads. A particularly difficult type of house to ventilate is the 'back-to-back' house where there is no through draught. The building of back-to-back houses is now prohibited, but many still exist in industrial areas.

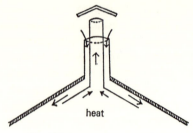

Fig. 138. McKinnel's ventilator.

(2) Mechanical ventilation. In many modern buildings such as cinemas and theatres, which accommodate large numbers of people – as many as 20 persons per 100 ft^2 – natural ventilation is not sufficient and recourse has to be made to mechanical methods. Mechanical ventilators may extract air from a building or propel air into a building – or do both.

In the extraction type of ventilation, air is drawn out of buildings by ducts situated at some distance from the places where the air enters; the suction force is provided by fans. By choosing the position of the exhaust ducts carefully a proper circulation of air is caused. This method is especially useful in factories where the manufacturing processes produce fumes or dust. The 'fume cupboard' of the laboratory is ventilated by this method.

In the propulsion system air is forced into the building. The advantage of this system is that the amount of air supplied may be regulated to the demand. The air may also be 'conditioned' – raised to a suitable temperature and moistened and cleaned if necessary. In some systems the air is warmed sufficiently to heat the building as well as ventilate it.

The most satisfactory systems of mechanical ventilation are those in which both propulsion and extraction are employed. By careful regulation the flow of air may be controlled and the sensation of draught avoided. The principles of mechanical ventilation are simple: in practice, however, ventilation has become a specialized branch of engineering, and the success of any system depends on the careful planning of the size and position of the numerous air ducts.

The most effective development in mechanical systems is that of 'air conditioning'. Provision can be made in these systems for heating, drying and humidifying the air; in special instances, e.g. in operating theatres or wards where patients are treated who are particularly liable to infection, the air may be sterilized. It is also possible to sterilize the air leaving a building when necessary, as in laboratories where work involves the use of highly infectious material.

There are several methods of testing the **efficiency of the ventilation** of a building. The best one is to visit a building at a time when it is crowded. The senses indicate whether ventilation is effective or not.

Inadequate ventilation produces effects such as lassitude, headache and a loss of a sense of 'well being'; if long continued, persons concerned become specially liable to respiratory infections. The risks of spread of droplet infections (see page 314) are greatly increased, and good ventilation is of particular importance during epidemics.

5 Methods of heating and lighting. Construction of buildings

Heating of buildings

Heat may pass from one body to another by one or more of three methods.

(1) **Radiation.** Radiant heat is transmitted in straight lines in the same manner as light, and does not heat the air through which it passes. The intensity of heat is inversely proportional to the square of the distance from the source. The sun's heat reaches the earth by radiation.

(2) **Conduction.** Heat may pass from one body to another by direct contact. Metals conduct heat easily; substances such as wood, felt and asbestos conduct heat only with difficulty and are known as 'insulators'.

(3) **Convection.** By this means heat is transferred by the actual movement of material. This form of heat distribution is only possible in liquid or gaseous media. Air, when heated by a warm surface, becomes lighter and rises, allowing heavier cool air to take its place; a similar result follows the warming of water.

The unit of heat used in this country is the amount of heat required to raise the temperature of 1 lb of water through 1 deg F (British Thermal Unit or Btu). The commercial unit in the gas industry is the Therm or 100,000 Btu. One electrical unit is equivalent to 3,412 Btu.

(N.B. The Btu is becoming obsolete with the adoption of S.I., but while still in use it must by definition depend on the degree Fahrenheit.)

Space heating

The heating of buildings involves the heating of the *space* of the building, e.g. rooms, corridors, passages, etc., and is known as 'space heating'.

Space heating depends on radiation, convection, or a combination of both. The generally accepted temperature for living rooms, offices, etc., is about 18°C.

Heat output required to obtain this temperature depends on the volume of the space to be heated, the outside temperature, and the insulating properties of the building – the material used in its construction. In a climate such as that of Great Britain, in general an output of three British thermal units per hour is required per cubic foot of space. As an example, the heating system for a small living-room would have to provide about 6,000 Btu per hour, the average school classroom about 30,000 Btu per hour, and a hospital ward of 20 beds some 80,000 Btu per hour.

An important factor in considering the merits of the different systems of heating is the 'efficiency' of the source of heat. Theoretically, it is possible to calculate the amount of heat produced by a given amount of fuel. In practice, however, the fuel may not be completely consumed or some of the heat produced may be wasted – that is, not applied to its main purpose. The percentage of heat which is actually available for the purpose for which it is intended, as compared with the theoretical amount of heat a given quantity of fuel can produce, is known as the 'efficiency' of the particular source. The following types of appliance are in general use:

(1) **Open fires.** The use of solid fuel open fires is still the most popular method of heating rooms. In addition to heating, open fires are a valuable aid to ventilation.

The old-fashioned open fire had a very low efficiency, but its design has undergone great improvement in recent years, and the newer types, heating by both convected warm air and radiation, are designed to burn smokeless fuel. The efficiency of the modern grate is 20 to 30 per cent.

(2) **Heating stoves,** which burn solid fuels, heat largely by convection, and give continuous and evenly distributed warmth. The type which opens, affording a view of the fire, has an efficiency of about 50 per cent, and the closed type about 70 per cent. Smokeless fuels are particularly suitable. Both open fires and solid fuel stoves can be provided with a water boiler.

Heating stoves can also use oil fuel, and types are now available which, for all practical purposes, obviate any oily smell. These stoves are usually transportable, with the advantage that they can be moved to where they are required; their disadvantage is that, being movable, they can be upset. Much attention has recently been paid to the design of these stoves to minimize any risk of fire due to their being upset, or due to the flame being affected by a draught.

Although the commonest type is the portable stove, oil fuel burners can also be used in fixed stoves and solid fuel stoves can be converted to use oil fuel by the installation of oil burners. The efficiency of oil stoves is of the order of 60 to 70 per cent.

(3) **The modern gas fire,** which is clean and convenient, has a radiation efficiency of about 50 per cent of the energy of combustion, and has a cheery appearance. It is designed to aid ventilation and must be provided with a flue.

(4) **Electric radiators** are extremely efficient, and do not produce fumes. Variations of the well-known electric fire are 'electric panels' built into the walls or ceiling. Electrical methods of heating are clean and convenient, but expensive. A modern development is to use 'heat storage' units which are heated during 'off peak' periods when electricity is available at cheaper rates.

(5) **Steam heating.** The above methods are those used for the heating of individual rooms. When large buildings have to be heated, the most efficient method is the use of central heating. The heat source is a central boiler, and the heat is distributed either as steam or as hot water round the building by means of pipes and radiators. The source of heat for the boiler may be either solid fuel, oil, gas or electricity. In large plants, if solid fuel is applied, it is now usual to introduce automatic stoking equipment.

The steam produced in a boiler is passed through pipes to radiators, the cooled steam being condensed to water and led back to the boiler. The steam is kept in circulation either by the pressure, which is regulated by valves, of the oncoming steam or the suction of the condensed steam. The disadvantage of this method is the high temperature to which pipes and radiators are raised. The method is useful for warming operating theatres, where a temperature of as high as 21°C may be required at short notice.

(6) **Hot-water heating.** Hot-water heating is the common method of central heating in use in this country. The hot water may circulate by the force of gravity or the flow may be aided by small 'booster' pumps driven by steam or electricity. The presence of a pump enables smaller

pipes to be employed, and in large buildings considerably reduces the cost of the installation.

In small houses the water is heated directly in the boiler: in larger institutions, however, it is found more effective to produce steam in a high-pressure boiler, and pass it through coils of pipes or 'calorifiers', where water is heated to the required temperature.

Heating by means of a calorifier is known as indirect heating, and is the usual method when a building has both central heating and requires ordinary hot water. Water for central heating has to be raised to a temperature of over 82°C, while domestic hot water has only to be raised to 60°C. By the use of a calorifier, the water for central heating can heat the domestic hot water. Only one source of heat is required, with consequent economy in fuel.

The water in the pipes may be under atmospheric pressure – 'on open system' or under more than atmospheric pressure – the 'closed or pressure' system. Higher temperatures may be reached in the pressure system, but the system is more complicated than the open system.

In houses the domestic hot-water system and the central heating system may be worked off the same boiler, valves cutting the heating system out of the circulation during warm weather.

Hot-water pipes are usually sufficient to heat small rooms; in large rooms and halls, radiators may be required.

(7) Heating by means of **warm air.** In this system air is passed over the surface of furnaces and carried by ducts to the various rooms. This method may be combined with the propulsion system of ventilation.

(8) **Panel heating.** A further development is to introduce the heating elements into the walls, floors or ceilings of the building itself. It is essential that this form of heating should be installed during the construction of the building.

For domestic purposes, panel heating is normally by electricity; in industrial premises it may be by means of steam or hot-water pipes embedded in the walls or ceilings. This method is efficient and produces an even heat throughout the room but care has to be taken to ensure that floors are not heated to a temperature that will cause discomfort.

In many methods of heating, it is now common practice to introduce a thermostat that can be set to keep the temperature at a predetermined level.

Lighting of buildings

The 'candle-power', the unit of intensity of a light source, was originally the light produced by a standard candle composed of sperm wax, and burning at the rate of 120 grains of spermaceti per hour. The 'standard light' is now maintained by incandescent electric lamps.

The 'foot-candle' is the unit of intensity of illumination, and is the illumination received by any object one foot from a source of light of one candle-power.

The methods of measuring the intensity of illumination are described in text-books on physics. Light obeys the inverse square law – the intensity varies inversely as the square of the distance from the source.

The brightness of an object depends on the amount of light it emits. This brightness may be due to the object emitting light itself or to the reflection of light. Light may be divided into two categories.

(a) Natural lighting

All natural lighting comes ultimately from the sun.

Its adequacy depends primarily on the brightness of any particular day, and the provision of sufficient windows. The colouring of walls and ceilings affects natural lighting, e.g. a bright room is lighter than one decorated in sombre colours. Whenever possible, natural lighting should be used in preference to artificial methods.

(b) Artificial lighting

Artificial light may be obtained from electricity, from different gases such as coal gas or acetylene gas, from paraffin oil, candles, etc. Electricity is the most efficient illuminant; it is clean, there are no products of combustion, and it is easy to carry electric current to any part of a building.

With the development of the national 'grid' for the distribution of electricity, it is now becoming available even to isolated houses, and is fast replacing other methods of lighting.

Artificial lighting may illuminate a room in the following ways:

(1) *Direct lighting*. Light passes directly on to an object or work-bench. The object is well illuminated, but glare and hard shadows, which are fatiguing to the eyes, are produced.

(2) *Indirect lighting*. In this method no light reaches an object directly. The whole of the light is reflected from the ceiling and walls, giving a diffuse light with no shadows. This method is expensive, as much light is lost in the process of reflection.

(3) *Semi-indirect lighting.* This is a combination of direct and indirect methods. The most effective method is to enclose the light in a diffusing glass bowl. Some of the light reaches an object directly and some of it is reflected from the ceiling. This method is not wasteful, and avoids glare.

Different intensities of light are required for different purposes. The following table gives a satisfactory intensity of illumination:

	Foot-candles
Hospital corridors	3
Hospital wards	3
Operating table	100
Operating theatre	10
Kitchen	8
Lounges	6
Reading rooms	5
General office	10

Industrial processes, as much as 50 foot-candles, if the work is fine.

The use of electrical 'fluorescent lighting' has become widespread. The ordinary electric bulb emits light because the electric current heats a fine coil of tungsten wire within the bulb, while in fluorescent lighting, the light is produced by the discharge of electric current heating the gas through which it passes.

Fluorescent tubes are considerably more expensive than electric bulbs; on the other hand, for a given amount of electric current, they produce much better illumination.

Building sites and materials

Site. A site on which buildings are to be constructed should be dry and warm. Sand, gravel and porous soils make good sites as they retain little water, which runs off them easily. A clay site is usually unsatisfactory, being damp and cold owing to its retaining much water. A damp site may be improved by subsoil drainage. 'Made-up' soil, i.e. soil made from the tipping of refuse, is never satisfactory.

A gentle slope and slight elevation are of advantage to a building site as they facilitate adequate drainage.

Aspect of a building. The aspect of a building is the direction in which the main rooms of the building face. In the northern hemisphere, rooms with a southerly aspect receive the maximum amount of sunlight. Living-rooms and bedrooms should therefore be placed on the south side of a house. South-east is probably preferable to due south, as the room receives the early morning sun and is protected from the heat of the afternoon sun. As the prevailing winds are from the south-west, a south-east aspect affords some protection from them.

The advent of the refrigerator has had a considerable effect on the design of domestic buildings. A former practice was, whenever possible, to place kitchens, food stores, etc., with a northerly aspect in order that the food should be protected from the heat of the sun and kept as cool as possible. This is not now essential, as the refrigerator can hold food at a suitable temperature (approximately 36°F), thus giving much greater freedom in the planning of buildings.

Buildings should not be crowded together on a site, as this results in lack of sunshine and fresh air. When land is scarce, as in cities, blocks of flats make the most effective use of the land available. Such blocks can be so sited as to provide adequate free space and so designed that the individual flats obtain the maximum of sunlight. Building upwards also saves agricultural land, but the cost of construction and maintenance is high.

Building materials. Houses may be constructed of brick, stone, concrete or slate, or wood. Roofs are constructed of tiles, slates or sometimes thatch. Since the Second World War, the shortage of 'traditional' building material such as bricks has given an impetus to newer methods of construction, e.g. reinforced concrete; the use of standard designs of houses has also enabled different parts to be mass-produced in factories (pre-fabrication), thus reducing the costs.

The weight of a building is carried on its foundations. The soil is excavated until a firm layer is reached and foundations are constructed of concrete or other firm material. The weight-bearing walls, usually constructed of brick, are built up on these foundations.

Until comparatively recently, non-weight-bearing walls were usually constructed of one row of brick covered by plaster, supported by wooden laths; ceilings were constructed of plaster also supported by laths. In recent years, however, there has been considerable development of new materials suitable for non-weight-bearing walls such as plasterboard, chipboard and others. Such materials are cheap and capable of being pre-fabricated in a factory, thus enabling the building to be more rapidly constructed. Particularly in large buildings, such as schools, hospitals, etc., the modern tendency is to make non-weight-bearing walls as readily movable as possible, thus enabling the internal partitions to be altered, and the use of the building to be made flexible.

Wood enters into the internal construction of houses, especially for floors. There should be no 'gaps' in which dirt can collect. Floors liable to become fouled and damp should be constructed of hard impermeable material. In living-rooms a pleasing effect may be obtained by a wooden block or parquet flooring.

Dampness is one of the difficulties in house construction. Dampness may reach the walls:

(1) By rising from the ground through the walls by capillary attraction. This is prevented by laying a horizontal layer of material impermeable to water in the wall just above ground level, or at the base of the exposed walls and chimneys. This layer, known as the 'damp-proof course', is composed of such material as asphalt, lead, slate, etc. The provision of damp-proof courses is now obligatory under the various building regulations.

(2) By infiltration of the walls, passing down exposed walls and chimneys. Such dampness may be avoided by the method of building 'cavity walls' – two separate walls built with a cavity of two inches in between them.

(3) By defective roofing, etc. This is avoided by careful workmanship. The risk may also be minimized by carrying the water away from roofs by gutters.

(4) By condensation. This form of dampness can be largely prevented by the careful selection of materials which have suitable heat-insulating properties.

Measures for the prevention of dampness are obviously more effective where they are introduced during the construction of a building. It is possible, however, although technically difficult, to introduce with modern equipment damp-proof courses into existing buildings. The insulating and heat retaining properties of buildings may be improved by the injection technique when a foam material is injected into cavities in the wall.

6 Water

The importance of water to human life will have been made clear in the chapters on physiology. Water is also necessary for hygienic purposes, such as washing, sewage disposal, street cleansing, etc., and is required in many trade processes. A plentiful supply of water is therefore an essential of present-day civilization.

The average consumption per head in a large city is surprisingly high but it must be remembered that those responsible for its supply must provide for industrial and municipal as well as domestic requirements. The average daily requirement of a town for domestic and municipal uses is approximately 30–35 gallons, the demands being made up as follows:

		gallons
(1) Domestic use	Drinking	0·33
	Cooking	0·75
	Washing	8·00
	Laundry	5·00
	Water closet	5·00
(2) Public Services	Street cleaning	5·00
	Public baths	
	Sewers	
	Fire-fighting	
(3) Trade purposes and unavoidable waste		8·00

In many industrial towns the consumption is higher and may reach as much as 60 gallons per day.

Sources and collection of water

The original source of all water is the sea. Water evaporates, condenses and reaches the earth as rain, finally returning to the sea by means of rivers, etc.

Rainwater. The amount of rain varies in different districts. In this country it is about 25 inches per annum in the east and reaches 80 inches in western districts.

Rainwater may be used for drinking water, but in practice not much use is made of it in this country, as roofs become foul and the rainwater becomes contaminated with soot, etc. When rainwater is used, it should be collected and stored in stone, well ventilated tanks, preferably placed underground.

Surface water. A proportion of water falling on the earth runs down the surface to form streams and rivers or collects in hollows forming ponds and lakes; it is known as surface water. An artificial lake or 'reservoir' may be made by damming a river or blocking the end of a valley through which a stream flows. Such a reservoir may be as large as a natural lake, e.g. Lake Vyrnwy in Wales, from which water for Liverpool is obtained, is over 10 miles in circumference.

Water flowing across the surface of populous districts collects much dirt, such as sewage and trade waste, and becomes quite unsuitable for drinking purposes. Water from sparsely populated hilly districts is free from such impurities and this 'upland surface water' forms one of the commonest sources of supply.

The area draining into a lake or reservoir is known as the 'catchment area', and should be as far as possible free from human habitation. The reservoir is known as the impounding reservoir, and usually holds water for at least 100 days for the area it supplies. From the impounding reservoir the water is conveyed by large rust-proof iron or concrete pipes to districts where it is required. External use is now being made of pipes constructed of plastic materials; such pipes are more easily handled than those of iron and concrete. Their length may be many miles – for example, Manchester takes its supply from the Lake District, a hundred miles away.

Water passes with or without purification to smaller reservoirs, usually constructed of concrete or brick, called 'service reservoirs'. The service reservoirs may be covered in and should be such a height above the area supplied as to ensure a pressure or 'head' of water sufficient to reach the highest buildings in the district.

Subsoil water. Some of the surface water soaks through the ground. The depth to which it goes varies with the geology of the country.

The earth is made up of various layers or 'strata'. Some of these strata allow water to pass through them and are termed 'permeable strata'. Other strata are 'impermeable' and water collects above them.

Geological strata do not remain at the same level under the surface of the earth. A deep stratum at one point may become the surface at another, forming an 'outcrop'. Any water which may be above this stratum comes to the surface at the outcrop, forming a spring; if there is no natural spring a well may be sunk to the level of an impermeable stratum.

Impermeable strata appear at various levels and water may collect above any of them; any stratum may form an outcrop. No stratum is complete and cracks allow water to leak through to a lower stratum.

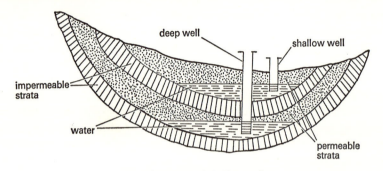

Fig. 139. Deep and shallow well.

Water may also reach a lower stratum from an outcrop (see Figs. 139 and 140).

A well or spring obtaining water from above the first impermeable stratum is known as a 'shallow' well or spring; when the well or spring taps water from below an impermeable stratum it is termed a 'deep' spring or well.

The terms 'deep' or 'shallow' give no indication of the actual depth of a well. A shallow well or spring is much affected by the rainfall and may dry up; it is also easily contaminated. Shallow wells are not good sources of water, but are still found in small villages and country districts. Deep wells form a good source of water and many cities in flat districts obtain their water in this way. If water from a deep well is under pressure and comes to the surface without pumping, the well is known as an 'Artesian well'. Fountains are sometimes supplied from Artesian wells.

Upland surface water and deep well or spring water form reasonably pure water supplies. Some towns have not such supplies available and have to obtain their water from rivers, etc., and purify it. The best known example of this is London. The Metropolitan Water Board obtains water

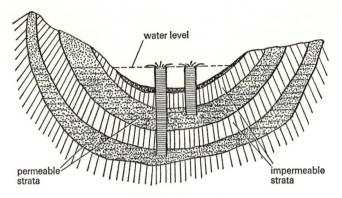

Fig. 140. Artesian wells.

from the Thames, which is heavily contaminated by sewage, etc., and purifies it by chemical and physical means.

Other sources. Fresh water may also be obtained by the removal of salt from sea water either by a distillation process or a freezing process. Such processes are technically difficult, but are being increasingly developed in parts of the world where there are no other available sources, e.g. desert areas. It is indicative of the steadily increasing demand for water that interest is now being taken in this source of supply in the United Kingdom.

Plants for removing salt from sea water are usually available in ships as an emergency stand-by.

Distribution to houses, etc.

Service reservoirs contain supplies of water for immediate use, and, as stated above, should give sufficient 'head' to enable water to reach above the highest buildings in the neighbourhood. In flat districts these reservoirs may have to be built in the form of water towers.

Water is carried to the various houses by pipes. There should be no chemical action between the water and pipe.

Water may be supplied to houses as:

(1) *An intermittent supply.* This means that water is only available for certain hours during the day, and necessitates storage in cisterns with risk of contamination. Pipes may become empty at some times, and if in any way defective, contaminating material may be sucked in. There is some slight reduction of waste in an intermittent supply.

(2) *A constant supply.* In this system water is available at all times, and

for drinking purposes no storage is necessary in the house, as water is drawn directly from the mains. Storage of water is only required for the hot-water system.

Impurities in water

Water which in nature is never chemically pure H_2O, contains many substances either in solution or in suspension. Some of these substances may improve the drinking qualities of a water, e.g. dissolved oxygen. The various impurities in water are acquired from its surroundings.

(1) *Rainwater* may be almost pure when it first condenses in the clouds, but soon becomes fouled by gases, soot and dust in the atmosphere. In industrial areas rainwater becomes heavily contaminated with industrial gaseous waste products such as sulphur dioxide, chlorine and ammonia. Further impurities are picked up from roofs, and the use of rainwater as a supply of drinking water is only practicable in country districts.

(2) *Surface water* and subsoil will naturally contain any of the impurities which may be present in rainwater. In addition it will gather impurities from the soil itself. These may be organic or inorganic in nature. Inorganic matter consists of salts dissolved from the soil, industrial waste, etc. Organic matter comes from sewage and decaying vegetable and animal matter.

(3) *Water from deep wells and springs* contains salts derived from the various strata through which it passes. By the time water reaches the deep strata many organic impurities will have been removed by filtration.

Substances in solution consist of inorganic chemicals from the soil and geological strata through which the water has passed. The dissolved chemicals may not have any deleterious effect on the drinking properties of water, e.g. lime salts may make the water 'hard', but it remains good for drinking purposes. Mineral springs consist of water in which there is a large amount of one particular type of salt, e.g. sulphur springs. Gases, such as oxygen and sulphur dioxide, may also be dissolved in water. Some gases give a 'live' taste to water. Water in which the oxygen has been driven off tastes 'flat'.

Suspended matter consists of small particles of decaying animal and vegetable matter, mud, clay, etc. This suspended matter is not, in itself, dangerous to humans, although it may be the cause of mild digestive upsets. The dangerous suspended matter consists of living organisms such as bacteria and protozoa, which are capable of causing such diseases as typhoid, dysentery, etc. Such dangerous types of contamination come from a human source, and the safety or otherwise of any water supply depends on whether human excreta reaches it or not.

The risk of contamination varies with different sources and the following is a list of sources in order of safety:

    Deep wells
    Deep springs
    Rainwater
    Large lakes – upland surface water
    Rivers
    Streams

Dangerous sources:
    Shallow wells
    Rivers – near banks
    Ponds

Protection of water supplies

Catchment areas are protected from contamination by strict control of dwellings and of grazing of animals in the area.

Wells may be protected by being 'steined' or lined with bricks set in cement. This ensures that water cannot percolate into the well without having filtered through a considerable distance of soil. Further protection can be afforded by fencing the area around the well and covering the mouth of the well when not in use.

Spring water may be protected by constructing covered reservoirs over the spring and fencing the nearby land.

Purification of water supplies

It is only very rarely that a community can obtain a water supply which does not require some form of purification.

The following main principles are employed in the purification of water:

Storage

Simple storage has a beneficial effect on the purity of water.

Suspended material tends to settle at the bottom, leaving the upper layers clear. This matter takes down with it many bacteria and other organisms. The process is a very slow one, but it can be expedited by

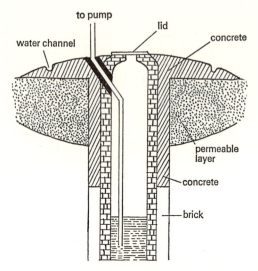

Fig. 141. Diagram of a well-protected well.

use of certain chemicals. The usual chemicals employed are salts of aluminium which form, with water, a jelly-like substance. This substance collects the suspended matter.

Another effect of the storage of water is that the disease-producing organisms disappear. Few of these organisms multiply in water under natural conditions and they therefore tend to die out.

Filtration

Practically all suspended matter may be removed by filtration through suitable substances. If the filter is fine enough, bacteria are also removed.

The action of a filter is a purely mechanical one. Its efficiency depends upon its fineness. It is not necessary to mention all types of filters. The commonest method in everyday practice is sand filtration. The principle is to run the water to be filtered over beds of fine sand and allow the water to soak through, when it is collected in pipes. The filtrative power of the sand is increased by the growth of an algal scum on it.

Water soaks through a sand filter-bed by the action of gravity. Mechanical sand filters have been developed in which the water is mechanically pumped through sand, but they do not differ in principle from gravity filter-beds.

Filtration may be employed for the purification of water on a small

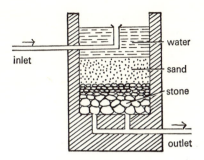

Fig. 142. A filter-bed.

scale for domestic purposes. Filters composed of such material as spongy iron, charcoal, sand, etc., either alone or in combination, do no more than clarify water and if not cleansed at frequent intervals become fouled with bacteria and actually cause an increase in the organisms in the water. An effective type of domestic filter is the Berkefeld or Pasteur Chamberlain 'Candle'. These filters consist of 'jars' of earthy material, such as porcelain or 'Kieselguhr' fitted into metal cases and so arranged that the water must pass through the porcelain. The 'mesh' of porcelain is small enough to arrest micro-organisms. If soundly constructed and frequently cleaned these filters can be safely relied upon to purify water. Their disadvantage is that the flow of water is slow.

Chemical sterilization

If a source of water is exceptionally heavily contaminated it may not be possible to render it safe by filtration and it becomes necessary to treat it chemically.

The principle of chemical sterilization is that all organic matter, including organisms, is destroyed by oxidation.

The disadvantages of chemical purification are that the substances used may taint the water; it is therefore necessary to remove the excess chemical substances.

Many oxidizing agents may be used. The usual one employed is chlorine. On a small scale, e.g. the Army water-cart, the chlorine is supplied to the water by means of bleaching powder; on a larger scale it may also be supplied by adding chlorine gas from cylinders to the water. A common process is to 'superchlorinate' the water. A high concentration of chlorine is added to the water and allowed to remain for a limited period – about twenty minutes. Excess chlorine is then neutralized by the addition of chemicals such as sulphur dioxide and sodium sulphate which also remove the unpleasant taste of chlorine.

For the sterilization of small quantities of water, it is now possible to obtain prepared sterilizing tablets in convenient form. All that is necessary is to add such tablets to water and wait, usually about half an hour. Such methods are very popular with caravanners and campers.

Chlorination is used on a very large scale in cities where the water comes from the heavily contaminated rivers.

Another method of chemical purification is the use of ozone (O_3). Ozonized air (produced by means of mains electricity) is forced through water in mixing chambers. The action is rapid and this method is used for sterilizing the water in swimming baths.

Other means of purification

Small quantities of water can always be made perfectly safe by boiling for at least 20 minutes.

It may also be sterilized by use of ultra-violet light, but this method is not as yet used much in practice.

Examination of water

The purity of a water supply demands continual vigilance by those responsible. In order to assist in this control both chemical and bacteriological tests have been developed. Such tests give an indication of the purity of the water and, if unsatisfactory, a clue to the source of contamination.

Waterborne diseases

The possibility of waterborne epidemics of infectious disease, particularly the enteric group, must be kept constantly in mind. Such epidemics used to be very common and it is probably true to say that some major outbreaks of waterborne disease in the early part of the nineteenth century did much to stimulate national interest in measures to improve sanitation and hygiene generally.

Waterborne epidemics are serious affairs. There is usually a period of about a month when a few cases arise followed by an explosive outbreak of a large number of cases. While modern water engineering and purification of water have made such outbreaks uncommon, the danger is always present and calls for constant vigilance on the part of the health authorities. A major outbreak of enteric fever in the United Kingdom in the 1930s and one in Switzerland in the 1960s are reminders of this ever-present danger.

Hardness of water

Some waters are said to be 'hard' because of their 'feel' when used for domestic purposes. This hardness is due to the presence of salts of magnesium and calcium and is of two kinds:

(1) Temporary (so-called because it may be removed by boiling), due to the bicarbonates of magnesium and calcium.

(2) Permanent, due to the sulphates of magnesium and calcium.

'Hardness' in water is not a severe danger to health. Such water has, however, the following disadvantages:

(1) It is wasteful of soap (1 lb of chalk may require as much as 10 lb of soap to produce a lather).

(2) Temporary hardness causes the formation of scales on boilers and pipes. This leads to waste of fuel and increases the danger of bursts.

(3) It may irritate delicate skins, and cause mild intestinal disturbances. These effects probably only occur in people unused to a hard water.

Removal of hardness

Temporary hardness may be removed by:

(1) **Boiling**

$$Ca(HCO_3)_2 \rightarrow CaCO_3 \text{ (precipitated)} + H_2O + CO_2$$

This method is not practicable on anything but a small scale.

(2) **The addition of lime**

$$Ca(HCO_3)_2 + Ca(OH)_2 = 2CaCO_3 + 2H_2O$$

Calcium carbonate being insoluble is precipitated, and may be removed by sedimentation or filtration through linen, etc. If the content of the water is known the approximate amount of lime can be calculated.

Permanent hardness may be removed by the **addition of soda.**

$$CaSO_4 + Na_2CO_3 = Na_2SO_4 + CaCO_3$$

The sodium sulphate is a harmless substance.

Combinations of silicate of aluminuim and soda are frequently used for domestic water softening, especially for permanent hardness.

The advantage of this system is that the efficiency of the chemicals may be restored by the use of salt solution, e.g.

$$Na_2Al_2Si_2O_8 + CaSO_4 = CaAl_2Si_2O_8 + Na_2SO_4$$

The chemical is restored by the 'action' of common salt solution:

$$CaAl_2Si_2O_8 + 2NaCl = Na_2Al_2Si_2O_8 + CaCl_2$$

This method is used on a large scale for removing permanent hardness.

The lime process is the more efficient in the treatment of temporary hardness, as both the precipitant and hardness are removed from solution and the total amount of solids in solution reduced.

7 Refuse disposal

To the animal body the removal of waste products is a necessity for health; in the same way the community must get rid of its waste products. If animal excreta, garbage, etc., are not regularly removed from human habitation the community becomes 'unhealthy' and outbreaks of disease occur.

The problem of the disposal of waste products is not a new one. The nomad tribes of Biblical days realized its importance and one finds in the Book of Deuteronomy (Chap. 23, vv. 12 and 13) instructions on this subject:

'Thou shalt have a place also without the camp, whither thou shalt go forth abroad;

'And thou shalt have a paddle upon thy weapon; and it shall be, when thou wilt ease thyself abroad, thou shalt dig therewith, and shalt turn back and cover that which cometh from thee.'

The simple instructions found in the Book of Deuteronomy were suitable for small and wandering communities, but, with modern civilization, the disposal of the large quantities of sewage and dry refuse has become a complicated problem.

Sewage disposal

Sewage includes human excreta, waste water, surface water, etc., and is usually disposed of by one of two systems.

Conservancy methods

In these methods excreta is retained in some kind of receptacle in or near a dwelling-house and periodically removed. Waste water must be dealt with separately and is usually allowed to soak into the land.

Conservancy methods are necessary in districts where the water supply is limited. For their successful working, excreta must be kept dry, and must not be mixed with waste household water.

One form of conservancy system is the 'privy' – a small chamber built of brick into which excetra is deposited; the chamber requires to be emptied at regular intervals. If refuse is also dumped into the privy a 'privy midden' is formed. In some houses the fixed receptacle of the privy is replaced by a movable pail – forming the pail closet. A more satisfactory method is the earth closet, where excreta when deposited is admixed with earth in galvanized iron containers. The earth acts as a deodorant. In all conservancy methods it is essential that the access of flies, etc., to the excreta is prevented, either by effective lids, preferably self-closing, or by covering with earth or similar material.

The excreta from the various receptacles may be dealt with either by digging into the ground or by incineration. Special vehicles may be required to remove the excreta if large quantities have to be dealt with.

In camps, excreta may be deposited directly into trenches. These trenches, which may be shallow if the camp is a temporary one, or deep in permanent camps, should be filled in when the camp is vacated, so as to minimize the risk of flies and dust spreading excremental disease (see Chapter 8). The name latrine is usually given to camp sanitary conveniences.

The modern development of the conservancy system is a chemical closet. Such closets are similar to the pail closet, the excreta being deposited in a chemical, usually based on cresol, which causes it to disintegrate and also acts as a deodorant. Arrangements can be made for appropriate ventilation, and such closets are quite satisfactory, even in confined spaces. They are very popular in caravans, cabin cruisers, etc.

Water-carriage system

In the water-carriage system excreta, waste water and surface water are removed from buildings through pipes to some place where they may be rendered harmless to health. A plentiful supply of water is essential for the proper working of this system.

The waste pipes from sanitary appliances (see later) unite outside a building to form a drain; drains join to form larger pipes or sewers, which carry the sewage to the site of disposal.

Drain-pipes, which are round or oval in shape, are constructed of glazed non-absorbent stoneware, or, if great strength is required, of

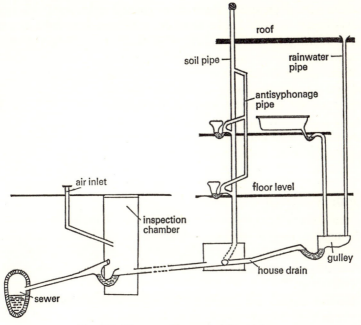

Fig. 143. Diagram of house-draining system.

cast iron. The inside surface is smooth to reduce to the minimum any resistance to the flow. The pipes are made in lengths of 2 ft to 3 ft with flanges to facilitate joints.

In any drainage system there must be a gradual downward slope from the drains to the sewage outfall, to ensure a steady flow of sewage. Bends tend to obstruct an even flow of sewage, and their number in any drainage system should be as few as possible. At each bend an inspection 'eye' – i.e. some means of access to the interior of the drain, should be provided for inspection and removal of any cause of obstruction.

Drainage systems require ventilation to prevent collection of foul air in the sewers. The temperature inside a drain is higher than that of the outside air; this causes air currents and an increase of air pressure in the system, which may unseal the water seals of traps (see later) and allow foul air to reach the dwelling-houses. The minimum requirements are ventilating shafts at the highest and lowest parts of a drainage system.

With the exception of the ventilating shafts all pipes entering a drain must be sealed to prevent foul gases being forced back from the drain into a dwelling-house. The usual type of seal is the simple 'water seal', and the part of the drain which forms the seal is known as the 'trap'. In its simplest form the trap consists of a pipe bent on itself (such a trap can be seen in the waste pipe of any wash-hand basin). The action of such a trap can best be illustrated by a simple diagram.

Water lodges in the pipe at the bend A, B, and prevents any gases passing through the trap. In such a trap the seal may be removed in numerous ways, e.g. evaporation, siphon effect, rush of water. Devices have been introduced to prevent this, but the principle of the trap remains unaltered.

Fig. 144. A trap.

Sanitary fittings

As the sanitary fittings in common domestic use are so well known, it will suffice to explain some more important points in their construction. They should be made of non-porous material such as fire-clay or vitreous china, and are designed to ensure that they are, as far as possible, self-cleansing.

Sanitary fittings are connected with the drain through a trap, the principle of which has already been described. In basins, baths, sinks, etc., the trap is formed by an S-bend in the outflow pipe. In the water closet the trap is incorporated in the closet itself.

Fittings used for excreta, e.g. water closets, slop sinks, etc., are known as **soil fittings.** They should be provided with a flushing cistern, usually constructed of cast iron and containing about two gallons of water. This tank is placed above the level of the closet, which causes a strong flush of water, and ensures a thorough cleansing of the pan of the closet after use. The flush is operated by siphonic action.

Soil fittings are connected directly to the drain by a 'soil-pipe', which is merely a continuation of the drain-pipe and seal. The soil-pipe may

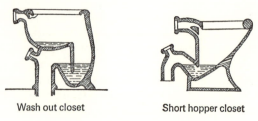

Wash out closet　　　　　Short hopper closet

Fig. 145. Two types of water closets.

be continued above the level of a water closet and act as a ventilating shaft for the drainage system (see Fig. 143).

Ordinary ablution and culinary fittings are connected to a 'waste pipe', which does not enter the drain directly but opens in the open air over a gulley, which is connected through a water seal to the drain. Rainwater pipes open over gullies in the same manner as do waste pipes.

Sewage

The large quantities of sewage collected by the water-carriage system has to be prevented from polluting streams and rivers; crude sewage would soon foul a river, destroying both animal and vegetable life and silting up the river-bed and causing a danger to health.

Discharge into the sea

This method is only available to towns at or near the sea coast. The sewage outfall should enter the sea below low-tide level. As currents are liable to wash sewage on to beaches the site of the outfall should be chosen so as to avoid this possibility.

Treatment of sewage

The aim of sewage treatment is to produce a final product or 'effluent' which is safe to discharge into rivers or allow to percolate through land.

Sewage disposal consists of the following different processes:

(1) **Removal of storm water.** In most sewage systems the pipes carry both sewage proper and storm water from roads, roofs, etc. After a period of dry weather, storm water may be as foul as sewage proper, and requires the same treatment. It is usual to treat three times the sewage produced in dry weather, i.e. 'dry weather flow'. Any storm water in excess of this amount is diverted from the sewage purification plant and passes without purification to rivers or land. The practice, however, of providing one system of drains and connecting waste water pipes directly to such drains is now becoming common. Rainwater is dealt with separately, thus avoiding overtaxing of the sewage works.

(2) **Removal of gross solids.** Gross material in sewage – solid material such as faeces, paper, etc., would soon clog up any sewage plant. To remove such material sewage is passed through an iron grid or screen with bars about $\frac{3}{4}$ in apart. The material caught may either be incinerated

or buried if suitable land is available, or ground up and returned to flow. Grit, etc., is removed by passing the screened sewage through tanks, detritus tanks, which slow down its flow and allow the grit to settle at the foot.

(3) **Removal of fine suspended matter.** The following methods are employed:

(i) *Sedimentation.* In this method the sewage is allowed either to flow very slowly through tanks or is run into tanks, where it is held for about four hours. The suspended matter gradually sinks to the bottom of the tank, forming 'sludge'. In the 'septic tank' sewage is retained in the tank rather longer than in ordinary tanks; the scum which forms over a septic tank encourages the action of organisms which thrive on the absence of free oxygen. These organisms are especially useful in decomposing organic matter. The septic tank is frequently used for the disposal of sewage at country houses or isolated institutions.

(ii) *Chemical precipitation.* In this method sedimentation is hastened by the addition of chemicals, lime and sulphate of iron being commonly employed. Chemical precipitation clears the sewage of suspended matter more effectively than ordinary sedimentation.

(4) **Disposal of sludge.** Sludge is the material which is deposited in sedimentation tanks or after chemical precipitation, and consists of about 90 per cent water. The solids of sludge are half organic and half inorganic.

Sludge may be disposed of by:

(a) Dumping in the sea.

(b) Dumping on land.

(c) Drying. It may then be burnt or used as manure.

(d) 'Digestion'. Some useful products such as marsh gas and fat may be extracted. This is followed by (a), (b) or (c).

(5) **Treatment of liquor from tanks.** The liquors from sedimentation tanks still contain organic matter in suspension and solution. This organic matter may be converted in the presence of free oxygen into harmless mineral salts by the action of bacteria and many other organisms. The oxidation is brought about by 'spreading' the liquor on as large a surface as possible. The methods used are:

(i) *Land irrigation.* The liquor filters through land. Light, porous soils are the most satisfactory for this purpose.

(ii) *Artificial biological methods.* In these methods the tank liquor is allowed to filter through clinker, where it comes in contact with the air.

The Sport and General Press Agency Ltd

1   Aerial view of the sewage treatment plant, Erith Kent (p. 303)

2   An insanitary well (p. 293)

3 Uncontrolled dumping of refuse which encourages flies, rats and other vermin (p. 206)

4 Rat guards (p. 322)

The process is aided by bacteria, fly larvae, etc., which stick to the clinker. Two modifications are used:

(*a*) Percolating filters. The clinker beds are usually round. The liquor is spread by means of rotary arms driven by the head of water.

(*b*) Contact beds. In this method the liquor is run into tanks filled with clinkers, where it remains for some hours. It is then run off.

The liquor obtained after treatment by the artificial biological methods – the effluent – may be run into rivers and streams without causing a nuisance. The effluent should not be allowed to contaminate water supplies.

The above description indicates the general principles of sewage disposal in communities. In isolated houses the septic tank is probably the most convenient method of disposal and the sludge must be removed periodically. The effluent may run into a ditch or be disposed of by land irrigation. Contamination of wells or springs, which may supply water to the house, may be avoided by careful siting of the tank.

One of the problems of sewage disposal which has recently become important is caused by the increasing use of synthetic detergents. They tend to produce a foam which is liable to be blown about by a wind and also tends to blanket the sewage from oxidation processes. When disposal is by discharge into the sea this foam is liable to be blown on to the beaches causing a nuisance and possible danger to health. Where foam is a problem, special sewage disposal methods are required.

Mention must also be made of the difficulties which arise through the use of radioactive substances in industry, hospitals, research institutes, etc. Such waste material has to be specially dealt with; before it can be introduced into the ordinary sewage system, the material has to be retained until the radioactivity has decayed to harmless levels, and it requires to be adequately diluted. The disposal of such waste is subject to stringent Government regulations.

Disposal of dry refuse

Dry refuse consists of household rubbish, such as ashes, papers, tins, bottles, etc., and some trade waste products. The amount of dry refuse is surprisingly large. A town of 50,000 inhabitants may produce almost 40 tons of refuse a day. The removal of refuse costs money and expense may be avoided if the householder or shopkeeper burns as much refuse as possible. Waste food should be burnt before it decays, or used as food for pigs, chickens, etc.

Refuse awaiting removal should be stored in galvanized iron receptacles provided with lids and so designed that they can be completely emptied. Fixed refuse receptacles or 'ash-pits' are never satisfactory; as they can never be completely emptied they encourage flies and rats.

Refuse bins, except in rural districts, are emptied at least once a week; in some areas a daily collection is made. All refuse vehicles should be provided with a sectional cover to prevent dust being blown about.

Refuse may be disposed of in the following manner:

(1) **Into the sea.** This method is only practicable in towns on the sea coast. Refuse is taken out to sea by special boats and should be taken far enough to prevent it being washed ashore.

(2) **Tipping on land.** Disused quarries, waste land of poor quality and low-lying land may be used. Haphazard tipping of refuse encourages flies, rats and other vermin. If tipping is done systematically it is a useful method and the 'made-up' land may be used for recreation grounds, parks, etc.

For tipping to be satisfactory the refuse must be laid in layers and covered with earth immediately. Each layer ought to be allowed to settle before another layer is laid. Tins, etc., must not be left exposed; large tins should be filled with ashes. This method is known as controlled tipping. The site selected ought to be some considerable distance from houses – the distance depends on the prevailing wind. Refuse should not be deposited in the collection of water often found in disused quarries.

(3) **Incineration and salvage.** In this method refuse is burnt in specially constructed incinerators. The heat produced may be used to produce steam which may be usefully employed. The 'clinker', the final result of incineration, is useful in road-making.

Tins, bottles, etc., cannot be incinerated and they may be separated and salvaged by hand or by passing the refuse through screens; the size of the mesh may be adjusted to separate tins or bottles, ashes, dust, etc. Salvaged material which has commercial value is sold and the proceeds of the sale reduce the cost of refuse disposal. Any other material is incinerated.

A modern method for the disposal of household refuse is the introduction in the outflow of the sink of a grinding mechanism by which refuse such as paper, vegetable leaves, potato peelings, etc., are ground up and disposed of in the waste water drains. In some methods, it is also possible to dispose of tins in this manner, although not glassware.

8 Micro-organisms and infection

Germs or micro-organisms are small unicellular organisms which play an important part in the economy of nature. The breakdown of organic matter in soil, essential for its fertilization, is a result of the activity of micro-organisms; many trade processes, such as brewing, depend on their presence.

A comparatively small group of organisms live on animal hosts. If they produce disease they are said to be 'pathogenic'. Organisms which produce disease in man are of interest in hygiene and medicine.

Micro-organisms, as their name indicates, are extremely small. A special unit – $\frac{1}{1000}$ millimetre, a micron or μ (pronounced 'mu') – has been introduced for their measurement. Most micro-organisms are of the order of 1 to 30 μ.

Individual organisms cannot be studied with the naked eye. They may be classified as:

(1) Bacteria
(2) Ultra-microscopic viruses
(3) Protozoa

Bacteria

The morphological characteristics of bacteria may be studied microscopically, and it is possible to identify and classify many bacteria by their appearance. Untreated preparations show the general shape of the organism and indicate whether it is motile or otherwise. Stained preparations show up the finer structure of bacteria. The usual stains are aniline dyes. As the manner in which the organism retains the dye varies, it is possible to classify organisms by their staining reaction, e.g. some retain the dye after treatment with alcohol or acid solution.

Certain organisms have two phases, an active phase and a quiescent phase. An organism in the quiescent phase is said to form a 'spore'. The feature of a spore is that it is not so easily killed as the active organism.

Bacteria may be cultivated outside the body. The material in or on which the bacteria grow, which may be solid or liquid, is known as the 'medium'. Artificial cultivation outside the body is known as cultivation *in vitro* (glass utensils being usually employed in bacteriology) in contrast

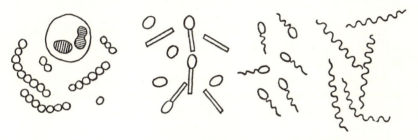

Fig. 146. Bacteria.

Left to right: streptococci (which are responsible for many kinds of inflammation); a white corpuscle is also shown. *Bacillus tetani* (which causes lock-jaw) with and without spores. *Nitrosomas* (which is found in the soil). A spirillum form.

to cultivation *in vivo* – in the living animal body. Animals may be employed to study organisms which normally affect man.

Single organisms multiply by fission, and if grown on a solid medium a colony is formed. The appearance of these colonies is characteristic of the organism.

Bacteria, for growth to occur, require moisture and a suitable temperature. Growth may take place between wide limits of temperature. The optimum temperature for human pathogens is about 37°C, the body temperature.

Some organisms require free oxygen for their growth – aerobes, while others only develop in the absence of free oxygen – anaerobes. The reaction of the medium also has some effect on the growth of organisms. The optimum reaction is generally one which is slightly alkaline.

Chemicals also have an effect on the growth of bacteria. Substances which inhibit the growth are termed 'antiseptics'; if the substance destroys the organism it is termed a disinfectant (see later).

Bacteria have a definite metabolism and they produce substances termed 'toxins'. A toxin is defined as the product which is injurious to the tissues of the host. The chemistry of toxins is not fully known. They are, however, known to be of a protein nature.

Toxins are of two kinds:

(1) *Exotoxins* or extracellular toxins, which diffuse from the bacteria into the surrounding medium.

(2) *Endotoxins* or intracellular toxins, which are retained within the cells until the bacteria disintegrate.

Exotoxins are unstable substances, easily destroyed by heat. They are specific for one organism, e.g. tetanus toxin causes muscular spasm or 'lock-jaw'.

Toxins are obtained artificially by cultivating the organism *in vitro* in a liquid medium. The organisms are removed by filtration through porcelain filters.

Endotoxins are more stable. They probably do not produce specific effects of disease, but are responsible for the general effects of toxaemia, which will be briefly noted later in this chapter.

The body reacts to:

(1) *All toxins*, by the formation of antitoxins which neutralize the toxin and render it harmless.

(2) *Bacteria*, by the formation of antibacterial substances. These substances bring about their effects in various ways. The main ones are:

    (*a*) Bacteriocidal: killing bacteria.

    (*b*) Opsonic: making the bacteria more susceptible to breakdown by the white blood corpuscles.

    (*c*) Agglutinating: causing the bacteria to clump.

    (*d*) Precipitating: causing the bacterial products to be precipitated.

Such reactions are termed *immunity reactions*, and the substances formed in the blood are specific for the invading organism.

The immunity produced by infection is termed naturally acquired immunity. Immunity may also be produced by artificial means.

(1) Injection of the dead organism, or a modified or attenuated organism which stimulates production of anti-bacterial substance, e.g. anti-typhoid inoculation.

(2) The injection of the toxin, or of the toxin slightly altered to form toxoid which stimulates production of specific anti-toxin, e.g. diphtheria immunization.

When the person himself produces the antibodies under the stimulus of toxin or bacteria, the immunity is termed an 'active immunity'. Such an immunity takes some time to develop but lasts for a considerable period.

(3) If the serum of an actively immune person is injected into a second person that second person acquires an immediate immunity due to the antibodies in the serum of the first person. Such an immunity is known as passive immunity and is developed very rapidly (within a few hours), but persists for only a short time – a week or two.

Both active immunity and passive immunity methods are used to control infectious diseases. Active immunization is most useful before a person is exposed to risk, as the immunity takes some time to develop. It is not so useful after a person has been exposed to infection – where passive immunization, which may be produced in a few hours, is the method usually employed.

Active immunity is produced by the inoculation of dead bacteria, known as 'vaccines', or in some diseases, of weakened or attenuated organisms. Passive immunity is produced by the injection of serum containing antibodies. This serum is prepared by the inoculation of animals – usually horses – with the organisms of the various diseases. Horses are not susceptible to many human diseases, and may be injected with large doses of the toxin or organisms of human disease without ill effect. It is thus possible to produce antitoxin in high concentration. After being obtained from the horse, the serum is further concentrated by chemical means.

Filterable or ultra-microscopic viruses

As already stated, ordinary bacteria may be studied microscopically and their size is of the order of one micron. Certain organisms are so small that they cannot be studied with the ordinary microscope but can be examined with the ultra-microscope using X-rays. These organisms are said to be ultra-microscopic, and called viruses.

Bacteria will not pass through fine filters and it is possible to free fluids from bacteria by filtration. The method of filtration used provides fluids free from bacteria as in the preparation of toxins. The ultra-microscopic viruses are small enough to pass through such filters and are also known as 'filterable viruses'.

The main characteristics of filterable virus are:

(1) The diseases produced are highly infectious from patient to patient.

(2) One attack normally produces a lasting immunity to further attacks by the same virus.

(3) Difficulty of cultivation *in vitro*, which calls for complicated technical processes of examination. The cultivation of the virus is only possible in living animal tissues, e.g. fertilized eggs or lungs of living mice.

Protozoa

These organisms are looked upon as the simplest kind of unicellular animals. The 'type' organism is the non-pathogenic Amoeba, which has already been described.

The protoplasm of protozoa is differentiated into nucleus and cytoplasm. The protozoa reproduce by fission and often exhibit a definite life cycle. Protozoal diseases are common in tropical countries, e.g. malaria (see Chapter 9).

Infection

When micro-organisms invade the body tissues, the result is disease, and the process is known as infection. The body's reaction to infection is complex, and the *symptoms and signs* are of two types:

(1) General symptoms – symptoms more or less common to all infections, known collectively as 'fever'. These symptoms are raised temperature, general malaise, vague pains, loss of appetite, vomiting, delirium, etc.

(2) Symptoms specific to the disease. Some infections give signs and symptoms peculiar to the disease, e.g. the typical rash in measles, signs of nervous disease in meningitis, local pain and swelling in wound infection, abdominal pain in typhoid fever, muscle spasm in lock-jaw, etc. The specific signs and symptoms enable a diagnosis to be made.

In a typical acute infectious disease such as scarlet fever or smallpox, the change from health to disease falls into more or less definite stages: This process may be recognized in other infections, although the stages are not so well marked.

(1) The incubation period. This period corresponds to the period of 'consolidation' of bacteria in the host. During the incubation period the person feels healthy, but may suffer from a few vague signs of ill-health.

(2) The prodromal period. The incubation period merges into the

prodromal period, which is marked by the patient showing the first general signs of infection. It is not usually possible to make a certain diagnosis during the prodromal period.

(3) The eruptive stage. In this stage the specific signs of the disease show themselves, and the illness reaches its height. The body's protective mechanism becomes mobilized and, if the patient is to recover, the illness passes to the next stage.

(4) The stage of recovery, in which the patient returns to health. The infection has been overcome.

When the infection is severe and overcomes the body resistance, death may result at any stage of the disease.

Micro-organisms produce their pathogenic action in three main ways:

(1) By toxins. The bacteria themselves remain at the site of entry, but their toxins enter the bloodstream and are carried over the whole body. The effect they produce varies with the disease, e.g. heart muscle is poisoned in diphtheria, a skin rash is produced in measles and scarlet fever.

(2) By local invasion of the tissues, e.g. wound infection, infection of the tonsils. There may be local destruction of the tissues. The tissues react by a process known as inflammation, by which the infection is walled off from the rest of the body. When the infection is completely walled off an abscess is formed. An abscess contains pus, a thick creamy fluid composed of bacteria, dead white blood corpuscles, and tissue débris. The formation of an abscess indicates that the defensive mechanism of the body has overcome the bacteria. If the body fails in this defence against the organism an abscess may not be formed and the organism may enter the bloodstream.

(3) The invasion of the bloodstream or generalized infection. The invasion occurs when the organism is of virulent type. The infection may start as a small infection, e.g. a prick of the finger; other portals of entry are the tonsils or the female genital tract after childbirth. This infection, if with virulent organisms, may never become localized, and organisms enter the bloodstream. The blood is very rich in protective substances or antibodies and the bacteria may fail to multiply but be destroyed by the antibodies. Such a condition is known as 'bacteriaemia'. When organisms multiply in the bloodstream the condition is known as 'septicaemia' or 'blood poisoning'. This condition is frequently fatal. Organisms which have reached the bloodstream may be carried to other parts of the body and produce multiple small abscesses. Such a condition is known as 'pyaemia'.

L

Healthy carriers

Some persons harbour disease germs without themselves suffering or having suffered from the disease. Such persons are known as healthy carriers and their importance in medicine is due to the fact that they are able to infect susceptible persons, who then suffer from the disease. Healthy carriers present, from the point of view of preventive medicine, a more difficult problem than actual cases of disease as their existence is often unsuspected.

Spread of infection

Infection may gain entrance to the body by various routes. A knowledge of the mode of entrance is of importance in nursing, as, if it is known, measures may be taken to minimize the risk of the spread of infection.

Organisms may be spread in several ways.

Droplet infection

During ordinary respiration a vast number of invisible droplets of moisture are exuded from the nose and mouth. Coughing and sneezing cause even more droplets to be 'fired' out into the surrounding air. A person sneezing can give another person a dose of droplets at a distance of as much as 10 ft. These droplets are liable to contain any organisms which happen to be in the throat. A person coming within 'range' of an infected person thus inspires the organisms of disease into the throat and may develop the disease.

The conditions spread by droplet infection are chiefly those of the respiratory tract, e.g. pneumonia, the common 'cold', influenza. The specific infectious diseases spread by this method include measles, diphtheria, scarlet fever, cerebro-spinal fever.

Preventive methods. A person is more likely to receive a dose of infective droplets the nearer he is to the sufferer. The risk of infection obeys the inverse square law and overcrowding and poor ventilation are predisposing factors to infection.

Preventive measures consist of:

(*a*) Good ventilation.

(*b*) Prevention of overcrowding. In dormitories, the greater the distance between the heads of sleepers the better.

(*c*) Nurses and doctors may obtain some protection by wearing masks over the nose and mouth when attending highly infectious conditions.

Infection through the alimentary tract

Organisms are ingested, and gain entrance by means of food, water or infected dust.

Excremental diseases usually have their effect on the alimentary canal, but they may cause more general diseases. Examples of such diseases are enteric fever, dysentery, cholera, infection by parasitic worms, enteritis, 'summer diarrhoea', etc. The organisms are excreted in the faeces and urine, and a human case or carrier is always the ultimate source of infection.

Organisms may be conveyed to food in many ways.

(a) By the hands of a food worker. Hands may become contaminated during defaecation, micturition, etc., if the food worker happens to be a carrier. Milk is commonly infected in this way.

(b) By infected dust. Dust is composed of small particles of street rubbish, etc. This may include animal dung, contents of badly kept sanitary closets, etc. Dust may settle on foodstuffs and thus convey the infection to other persons. Dust is especially dangerous in hot countries, where sanitary conditions are primitive.

(c) By means of flies. They breed and feed on animal excreta and also on food. It is thus an easy matter for infective material to be carried to human food. Such opportunities of infection are most common in camps, etc., where sanitation and the protection of food cannot be so easy to maintain as under ordinary conditions.

(d) Water is a frequent cause of outbreak of excremental disease. As water is universally drunk, water-borne epidemics are characterized by the more or less simultaneous occurrence of a large number of cases.

It must be remembered that the source of any infection is always a human case or carrier. Water supplies are contaminated by the careless person who defaecates or micturates at random or by the contamination of a source of water by sewage.

Prevention of water-borne infection depends on:

(a) Efficient disposal of sewage or human excreta.

(b) Protection of water supply from all possible, especially human, sources of infection.

The introduction of the water-carriage system of sewage disposal and the elimination of shallow wells as sources of water supply have reduced the incidence of the most typical excremental disease, typhoid fever, from 371 per million in 1871 to 5 to 6 per million in 1936.

Infected water can always be rendered safe by boiling for a sufficient time. This should be done whenever water is suspected, and a small amount of safe water is required.

Contact diseases

Such diseases are skin diseases. For their spread direct bodily contact is necessary between sufferer or a fomite and new case.

The prevention of the direct-contact disease depends on cleanliness and personal hygiene.

Venereal diseases are also contact diseases for they are transmitted from one person to another during sexual intercourse. The commonest, gonorrhoea, caused by a bacterium, and syphilis, caused by a protozoan, are both serious, but if diagnosed in the early stages they can be treated.

Prevention depends on restraining infected persons from having sexual relations with others.

Diseases requiring an intermediate host for their transmission

Certain diseases require the intervention of some intermediate host for their transmission from case to case. The feature of these diseases is that the intermediate host is essential for their spread; the animal does not merely act as a passive carrier of the infecting organism, as the housefly does in enteric infection.

Insects are the commonest 'vectors' or carriers of the disease. They infect the human being in different ways.

(*a*) By a bite, e.g. mosquitoes in malaria. In this case the organism goes through a definite phase of its life cycle on the insect. Bubonic plague is spread by the bite of an infected flea.

(*b*) By the inoculation of the excreta of an infected insect into small abrasions, e.g. louse in typhus.

Diseases of mammals conveyed to man

Examples are:

(*a*) *Rabies or hydrophobia.* Infection is spread by the bite of an infected dog, the virus being found in the saliva.

(*b*) *Undulant fever or Malta fever.* The organism is found in the milk of the goat or cow.

(*c*) *Spirochaetal jaundice.* This organism is found in the urine of rats, and can penetrate the human skin.

Disinfection

Disinfection is the process by which disease germs are killed and rendered harmless. The aim of disinfection is to break the chain of infection. With an increasing knowledge of the mode of spread of disease it is possible to use methods which attack the germs at their source, and at some definite point in their transmission. The old idea of disinfection aimed at the haphazard destruction of germs in the air and sick-room is not a very effective method.

The rational use of disinfectant is illustrated in enteric fever. The faeces are treated so as to destroy the enteric organisms – an attack on the source of the organisms. The nurse's hands are washed with some disinfectant – an attack on the method of spread. 'Fomites', e.g. the linen used by a sick person, are treated with disinfectant – another attack on the method of spread.

In the treatment of infectious disease the disinfection can be carried out as:

(1) Concurrent disinfection, i.e. disinfection of faeces, hands, etc., while the patient is ill. This is rational disinfection.

(2) Terminal disinfection, i.e. disinfection of room, bedclothes, etc., after patient has recovered. This is the old, rather haphazard attempt to destroy organisms which may be in the atmosphere.

Disinfection methods fall into two main groups, physical and chemical:

(A) Physical agents

Light. Ultra-violet rays kill bacteria. This method is not used extensively except perhaps in hot countries, where the sunlight, which includes light and ultra-violet rays, is strong. In this country all it is possible to say is that well lit rooms are more germ-free than badly lit rooms. As a well lit room is usually well ventilated, much emphasis cannot be placed on the efficiency of temperate sunlight.

Heat. Fire has always been looked upon from time immemorial as the 'great purifier'. Burning an article will always destroy infection, but such a course is somewhat drastic.

Heat may be applied in two forms:

(*a*) *Dry heat.* A temperature of 150°C for an hour destroys all forms of life, including bacterial spores. Most delicate fabrics, however, are destroyed at this temperature, and dry heat is not much used for disinfection. It is useful for 'disinfestation', i.e. the destruction of vermin. Lice and their eggs are, for example, destroyed at 60°C. A handy way of destroying such vermin is by the use of an ordinary iron. Dry heat may be used for sterilizing glass instruments in bacteriological work.

(b) *Moist heat.* Boiling will kill all disease germs. To kill the spores it must be sufficiently prolonged, or repeated after time has been given for the spores to germinate. It can be used for the disinfection of curtains, crockery, towels, etc. Steam is a useful disinfecting agent. It is reliable and penetrates fabric, etc. It ruins such materials as leather and rubber. Steam may be used as:

(i) Current steam, i.e. the steam passes through some kind of chamber at atmospheric pressure. Its temperature will be 100°C. It penetrates the materials and heats them by its latent heat.

(ii) Pressure steam. In this method steam is passed into a closed chamber under pressure. The value of such steam is that its temperature is higher, e.g. at 3 atmospheres (45 lb in²) its temperature reaches 144°C. Many complicated disinfecting machines have been designed, based on the employment of pressure steam. This method is used in the disinfecting plant of most hospitals.

In hospital practice, it is necessary to render surgical instruments, fabrics, etc., used in the treatment of patients, at operations for example, completely free from organisms. Such a process is known as 'sterilization', and is essentially the same as disinfection.

Modern sterilizers have a large capacity, and can sterilize rapidly. They are expensive and complicated machines, and in most hospitals sterilization is now carried out in central sterilization service departments. The establishment of such departments has led to the practice of supplying wards and operating theatres with the required instruments, fabrics, etc., for different procedures in sterilized packs. Such an arrangement saves considerable work in the ward.

Another development in this field is the use of the disposable instruments, syringes, etc., intended to be used once and then discarded. These items are supplied already sterilized by the manufacturers. Their use ensures a high standard of sterilization and saves considerable work in the hospitals themselves; these disposables, however, are somewhat expensive to use.

(B) Chemical agents

Chemical agents used for disinfection purposes may be gases or liquids. A distinction must be made between substances which destroy the organisms – disinfectants – and substances which merely inhibit their development – antiseptics. In many instances antiseptics are weak solutions of disinfectants.

Gases are most commonly employed in the disinfection of rooms, and their effect depends on a suitable temperature and as high a concentration as possible.

Examples of gases used:

(a) *Formaldehyde.* This gas is formed by the action of an oxidizing agent, such as bleaching powder on formalin. The gas is useful for disinfection but has little effect on insects.

(b) *Hydrocyanic acid gas.* Although very poisonous has not much effect on bacteria. Its main use is for the destruction of rats and bed-bugs.

(c) *Sulphur dioxide.* Obtained by burning sulphur. It has only a weak disinfectant action. It is an ancient method, 'fire and brimstone' being used in former days to destroy 'vapours' of disease. Its use nowadays has become restricted, but it is still used for destroying rats in ships.

Gases are not very effective disinfectants. They can only be used in 'terminal disinfection' and are an example of an empirical attempt to kill organisms in the air. They do not attack the source or route of infection of an organism.

Liquids. The largest group of liquid disinfectants is composed of coal-tar derivatives. The standard disinfectant is carbolic acid or phenol, and the strength of any new disinfectant is expressed in relation to phenol.

The commonest phenol derivatives are cresol and lysol. They can be used in weak solution for cleansing rooms. In stronger solution they can be used for disinfecting faeces, vomit, etc.

Mercury salts also act as disinfectants for such purposes as washing hands. They are very poisonous and should be used with care.

Bleaching powder is used as a disinfectant in strong solution. It can also be used as a deodorant.

Liquid disinfectants can be more easily applied to the source and route of infection than gases, and are therefore more effective in their actions than gases.

(C) Specific drugs

Until recently the treatment of infectious diseases depended largely on good nursing and general medical care designed to conserve the strength of the patient. In some diseases, notably diphtheria, a specific antitoxin serum was used. The advent of new synthetic drugs has provided physicians with means of attacking the infection itself. Such treatment is known as 'chemotherapy'. The antibiotic drugs, which are produced from bacteria themselves, and of which penicillin is the best known, also enable the physician to attack the source of infection directly. These drugs, given in the early stages of infectious disease, may abort or modify the attack, and they can thus be regarded as a form of prevention.

9 Personal parasites and the housefly

A **parasite** is an organism which obtains its nourishment from another organism, the host. It may live on the outside of the host (external parasite), or within the host's body (internal parasite).

External parasites

Parasites which live on the surface of the human body include: the flea, bed-bug, and three kinds of lice. They are unpleasant, but are not dangerous in themselves. They may, however, convey diseases from one person to another, e.g. the flea may spread plague, and the louse, typhus.

The flea

Fleas are flat wingless insects, which are capable of jumping long distances. They feed by piercing the skin and sucking the blood of their host. The bite causes a great deal of irritation.

The female flea lays its eggs in crevices in a room. The eggs hatch into larvae in from 5–12 days according to the temperature and humidity. The larvae feed on the organic matter in the dust and change into pupae in from 12–14 days. About 14 days later the adult insects emerge and begin biting their hosts.

Fig. 147. The flea (×15).

If undisturbed a young flea may live without food for four months. Fleas may therefore hatch and develop in empty houses.

Pulex irritans, the flea which lives on humans, transmits no disease, but *Xenopsylla cheopsis*, the rat flea, transmits plague and a form of typhus. If a rat flea sucks the blood of a plague-infected rat, the bacilli get into its stomach and multiply. When the rat dies from plague, the flea leaves and seeks out a fresh host. When it pierces the skin of this fresh host the flea infects it with plague. The fresh host may be any warm-blooded animal, including a human being.

It is interesting to notice that all outbreaks of plague in human beings have developed after an outbreak of plague in rats. Great care is taken at ports to prevent rats coming ashore from ships which may have come from plague-infected parts. Plate 4 shows special guards used to prevent rats coming ashore along ropes.

The breeding of fleas may be prevented by keeping the floors of a house scrupulously clean and by blocking up all crevices. If fleas do get into a house they may be destroyed by fumigating the room with sulphur dioxide or paradichlorobenzene and washing the floor and furniture with a paraffin emulsion and soft soap. Clothes should be treated with D.D.T. powder and washed.

Ships which have become infected with plague are fumigated with hydrocyanic acid which is lethal to rats.

(Cimex lectulrius) or bed-bug

Bed-bugs are flat, active insects, living chiefly in broken-down walls and floors. They only feed on their victims at night. This they do about every seven days, but may exist for a year without food. As a result they may be conveyed from one dwelling to another in second-hand furniture, etc. They are able to pass along gutters and walls from house to

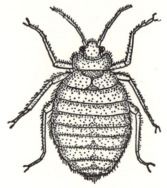

Fig. 148. The bed-bug ($\times 8$).

house. Their presence in a house is easily detected by their strong odour.

The eggs are laid in crevices and holes in the walls, joints of furniture, etc. They hatch out in from seven to twenty-one days, and are ready to begin their biting immediately.

Bugs are difficult to eliminate once they have got into a dwelling. Rooms should be fumigated with sulphur, or hydrocyanic acid, a poisonous gas which should be treated with the greatest of care and only used by a qualified specialist. Furniture removed from slums to new houses should be fumigated and treated with an insecticide such as D.D.T. in kerosene.

Lice

Lice bite the skin, producing inflamed patches and causing a great deal of irritation which may lead to impetigo. Throughout their life they seldom leave the body of a host.

The female louse secretes a kind of cement which attaches the eggs, commonly known as nits, to the hair or clothing. In about a week the eggs hatch into larvae which feed every 24 hours and moult three times before becoming adult. The adult, like the larvae, feeds at frequent intervals and is spread by contact of body and clothes.

There are three species of lice:

Pediculus capitis (or **head louse**). This louse is found on the hair of the head and bites the scalp. It is easily transferred from one individual to another.

The female is about $\frac{1}{8}$ in long, the male slightly less.

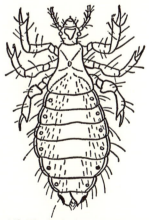

Fig. 149. The body louse ($\times 14$).

Pediculus corporis (**or body louse**). This louse is larger than the head louse and is found on the body or clothing. The female attaches its eggs to the fibre of the clothing.

Pediculus corporis may be responsible for conveying certain diseases including typhus fever, relapsing fever and trench fever.

Pediculus pubis (**or crab louse**). This louse lives in the hair covering the pubis. Owing to its small size it is difficult to detect.

Lice may be removed from the body by bathing. When the head is infected with *Pediculus capitis*, the hair should be rubbed with special hair oil which kills the adults and nits. This does not remove the nits and they must be cut out or loosened by soaking the hair in vinegar for about half an hour. The hair should then be combed with a small-tooth comb and washed. An emulsion of benzyl benzoate will both kill the lice and nits and dissolve the cement, but this substance is expensive.

Clothes infected with lice should be well dusted with D.D.T. powder and washed.

Sarcoptes scabii (The itch-mite)

The itch-mite is the cause of scabies, a contagious skin disease. The female itch-mite burrows into the skin, especially in creases of joints and between fingers, and produces great irritation. In the hole formed, it lays about twenty-four eggs and then dies. The eggs hatch out into larvae. These in two or three days burrow deeper and form blisters or vesicles which characterize this skin disease.

In about three weeks the mites are fully grown and they come to the surface. There they mate, and the life cycle is repeated, the male mites dying off after mating has taken place. Each female mite dies after a very short life, though it can exist longer if no suitable breeding ground is found immediately. For this reason clothing may be infected for a long time.

The presence of the itch-mite on the body is caused by contact with the clothes or body of an infected person. It is encouraged by lack of cleanliness.

To cure scabies the skin should be painted with a lotion containing benzyl benzoate, soft soap and boric acid. The clothes and bedding should be disinfected with steam.

Internal parasites

Bacteria and protozoa, e.g. malarial parasite, are internal parasites;

roundworms (nematodes) and flatworms (platyhelminthes) may also live parasitically inside the human body.

Parasitic nematodes

In temperate countries *Ascaris lumbricoides* and *Enterobius vermicularis* are the commonest nematodes parasitic in the bodies of human beings.

Ascaris lumbricoides **(or human roundworm).** The adult worm lives in the small intestine, chiefly of children. It is whitish in colour and not segmented.

There are two sexes; the female is about 10 in long, the male about 8 in. The worm has a mouth guarded by three lips, through which it absorbs digested food from the intestine of the host.

The eggs are fertilized within the body of the female. They pass into the intestine and then out of the host in the faeces. They may infect napkins, clothing, water supply, etc. They often get lodged in a child's finger-nails.

Outside the gut of the host they undergo a period of ripening, for which they require a temperature of 15·5°C, plenty of moisture and oxygen.

Re-infection takes place through the mouth. The egg reaches the intestine and hatches into a larva. Before a larva develops into an adult worm it wanders round the body. It pierces the intestinal epithelium and passes into the blood. It goes via the blood to the heart and then the lungs. There it pierces the lung walls and enters the cavity of the lungs. It then passes up the trachea and enters the alimentary canal. When it reaches the intestine it grows into the adult worm and the life history is repeated.

Normally the adult worm causes little trouble to the host, though sometimes it is the source of diarrhoea, anaemia, and nervous complaints, especially in children. The larva during its wanderings may cause inflammation of the lungs.

Enterobius vermicularis **(or threadworm).** This worm is also found chiefly in children. It is much smaller than the roundworm (the male and female are both less than half an inch), and it inhabits the caecum and rectum. It causes anal irritation. As in *Ascaris* re-infection takes place through the mouth. In *Enterobius* there is no larval wandering.

Parasitic platyhelminthes

Parasitic flatworms include the tapeworms and the flukes. In this country there are no flukes parasitic in human beings.

Taenia solium. This tapeworm in the adult form is parasitic in the small intestine of man. It is yellowish in colour and is about ten feet long. At the front end there is a small head about the size of a pin-head. This head has a ring of hooks and four suckers. The hooks and suckers enable the parasite to cling on to the intestinal epithelium.

Behind the head there is a thin neck and then numerous segments called the proglottides. New proglottides are continually cut off from the neck and they grow rapidly. The worm gets wider, farther away from the head.

The tapeworm is monoecious and within each segment or proglottis there is a complete set of male and female reproductive organs. When the

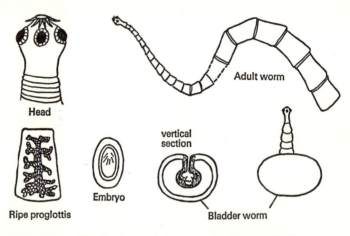

Fig. 150. Life history of the tapeworm (*Taenia solium*).

proglottides become full of fertilized eggs they break off from the rest of the worm and pass out of the body in the faeces. The eggs begin to develop into embryos.

In the soil the wall of each proglottis decays and ripe embryos covered with a firm resistant egg-shell are set free. Each embryo consists of a mass of cells with three pairs of hooks.

No further development can take place until the embryo is eaten by a pig. Inside the gut of a pig the egg-shell is dissolved and the six-hooked embryo bores its way through the wall of the gut into the blood vessels. It travels in the blood to the muscles, where it becomes embedded. The surrounding muscular tissue swells up and a cyst is formed. The embryo looses its hooks and develops into a bladder worm (see Fig. 150).

No further development can take place unless this 'measly pork', in an undercooked condition, is eaten by man. Inside the gut of man the head is thrust out of the bladder and begins clinging to the gut wall. The bladder shrivels, and the head begins cutting off proglottides. The life history of the animal is then repeated.

Other tapeworms have life histories essentially the same as *Taenia solium*, though their hosts may vary. Some of the common tapeworms are compared in the table below:

Tapeworm	No. of segments	Head	Host of adult	Host of bladder worm
T. solium	1,000	Hooks and suckers	Man	Pig
T. dibothriocephalus	3,000	Only suckers	Man	Fish
T. saginata	2,000	Only suckers	Man	Ox
T. echinococcus	4 (incl. head)	Hooks and suckers	Dog	Man, sheep

T. echinococcus in man is only found in countries where men live in close association with dogs.

The housefly

The housefly, though not a parasite, plays an important part in the distribution of disease and will be described in this chapter.

The female fly lays her eggs throughout the year in moist fermenting animal or vegetable matter. These eggs which are laid in masses are small (about 2 mm long) white, cigar-shaped structures. The time they take to hatch depends on the outside temperature and varies from 2 to 14 days. The maggots come out, feed on the refuse and grow very rapidly, becoming $\frac{1}{2}$ in long in four days. They then pupate and the adult fly hatches out from the pupa in about a week.

The housefly lives on filth and human food indiscriminately, and may carry germs from the filth to the food in several ways:

(1) The fly has hairy legs with sticky pads at the end. Filth may be conveyed from place to place on these pads.

(2) The fly eats filth and the germs in it may survive in the stomach for several days. When the fly alights on food it may vomit up the contents of its stomach, or the germs may pass out of the fly in the faeces.

Measures against flies

(1) Food should be kept in a larder with the windows covered with some perforated material to prevent flies entering.

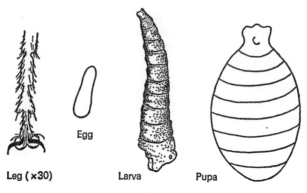

Fig. 151. The housefly.

(2) All foodstuffs, especially milk, should be covered.

(3) Flies should not be allowed to breed. All refuse should be burnt, or if this is not possible, sprayed with kerosene to kill the maggots.

(4) Dustbins should have tightly fitting lids.

(5) A refuse dump should be placed away from the house.

(6) Tips should be covered with earth.

Mosquito

There are two types of mosquito – the Anophelines and the Culicines (gnats). The Anophelines are responsible for the spread of malaria; the Culicines are generally harmless. Fig. 152 shows the two types of mosquito in resting position.

Part of the life history of the mosquito is spent in water. The female mosquito lays its eggs in clumps on the surface of stagnant or semi-stagnant water. The larvae hatch out in two to three days. They are very active and move about in the water, coming to the surface to take in air through a tube near the end of the body.

In about ten days, after moulting several times, the larvae develop into

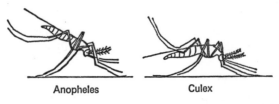

Fig. 152. Anopheline mosquito (left), Culicine mosquito (right).

active comma-shaped pupae. These also come to the surface to breathe, taking air in through two tubes near the head end.

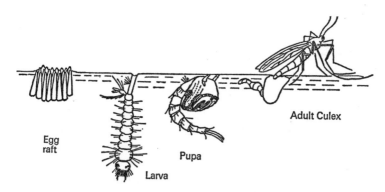

Fig. 153. Life history of a Culex mosquito.

The outer coverings of the pupae split in about seven days and the winged insects emerge.

The male mosquito does not feed when adult; the female mosquito pierces the skin of mammals with its long proboscis and draws blood.

Measures against mosquitoes

(1) The breeding-places of the mosquito, such as marshes, stagnant ponds, etc., should be removed as far as possible by draining.

(2) If (1) is not possible, the water should be sprayed with oil. The larvae and pupae are poisoned by substances in the oil.

(3) Fish which feed on the mosquito larvae should be encouraged to breed.

(4) In countries where malaria is prevalent the windows and doors should be covered with netting to prevent the entrance of mosquitoes into houses, and nets should be placed over beds.

Malaria

When the female mosquito sucks the blood of a person infected with the malarial parasite, a unicellular animal similar to Amoeba, some malarial cells pass into its body. There they go through the complicated sexual stage and multiply profusely.

The malaria cells eventually reach the salivary glands of the mosquito, and are injected into the body of the mosquito's next victim. In the blood of a human being the malaria cells go through an asexual cycle every two or three days, the end of each cycle producing a rise in temperature in the patient.

Insecticides

Insect pests can be controlled by 'insecticides', substances that are lethal to insects. From the point of view of their action they may be divided into two groups:

(1) **Non-residual insecticides,** a typical example of which is pyrethrum, a vegetable product. The feature of this group of insecticides is that they have a quick effect – an immediate 'knock-down'. They are particularly useful in the form of a spray to deal with flying insects.

(2) **Residual insecticides.** These are synthetic chemicals and the effect is a persistent one. The eggs and larvae of insects are more resistant than the adult insect, and the persistent insecticide is very valuable in dealing with the adults as they hatch out.

Persistent insecticides do not have a quick 'knock-down' effect, and in practice insecticide preparations are a mixture of both non-residual and residual insecticides. They are available as solutions, sprays, dusting powder, etc., and may be combined with paint or distemper.

The introduction of persistent insecticides made possible great advances in insect control. The best known is D.D.T. (dichloro-diphenyl-trichloro-ethane) which was extensively used during the Second World War. Applied for example as a dusting powder, it was possible to deal effectively with large numbers of lice-infected persons, and thus control outbreaks of typhus fever.

There are a large number of synthetic insecticides. They, like D.D.T., are chlorinated hydrocarbon derivatives, and others are based on phosphorous compounds.

Skill and care is necessary in the use of insecticides as many are toxic, demanding the provision of protection for those working with them. It should also be noted that insects can develop a resistance to particular insecticides; also that vegetation may be affected by them.

PRACTICAL

1 Examine with a magnifying-glass prepared slides of flea, bed-bug, itch-mite and the three species of louse.

2 Examine a preserved roundworm and threadworm.

3 Examine under the microscope prepared slides of the different stages of tapeworm.

4 Examine larvae and pupae of housefly. Examine the winged insect. Mount one of its legs and examine under the microscope.

5 Examine under the microscope a prepared slide showing stages of the malaria parasite. (An oil immersion lens should be used.)

6 Observe in an aquarium the development of a gnat from the egg stage to the adult insect.

7 Examine under a magnifying-glass a true mosquito and a gnat.

10 Principles of hygiene applied to hospitals

Since the days of the Crimean War, improvements in the hygiene of wards where sick people are nursed have gone hand in hand with improvements in the art of nursing. Florence Nightingale was a pioneer not only in nursing, but in improving the fundamental conditions under which the work is carried out. Her *Notes on Nursing*, written in 1859, prove this. For example:

(1) The very first rule of nursing . . . 'to keep the air the patient breathes as pure as the external air, without chilling him.'

(2) 'The life duration of babies is the most delicate test of sanitary conditions.'

(3) 'The air throughout a room is never changed by a draught in the lower part of the room; but it *is* changed by an open window in the upper part.'

(4) 'The safest atmosphere for a patient is a good fire and an open window.'

(5) 'People think that the effect (of the sun) is upon the spirits only . . . Light has quite as real effects upon the human body.'

(6) 'All house drains should begin and end outside the walls.'

These excerpts at the present time seem trite and commonplace, but it must be remembered that in those days there was no science of hygiene as we know it today, and the soldiers and civilians were nursed under appalling conditions. Four-poster beds with two or more patients in them were not unknown and, of course, the germ theory of disease was yet to be discovered. The work of Pasteur in the 1870s again made alterations in hospital hygiene imperative, and gradually the hospital ward of today has evolved.

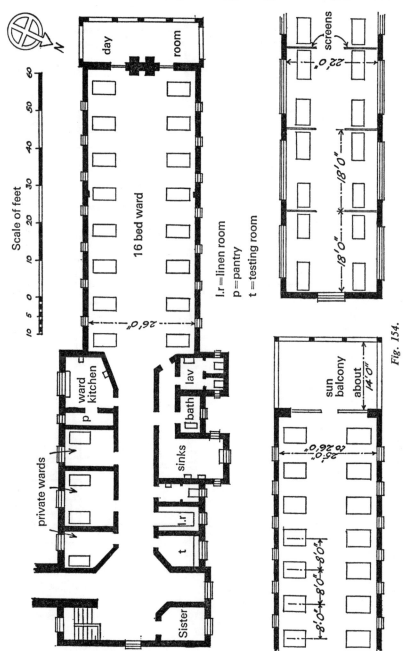

Scale of feet

day room

16 bed ward

26'0"

l.r = linen room
p = pantry
t = testing room

ward kitchen

p

private wards

sinks

bath

lav

l.r

t

Sister

screens

22'0"

18'0"

18'0"

sun balcony

about 14'0"

25'0" to 26'0"

8'0" 8'0" 8'0"

Fig. 154.

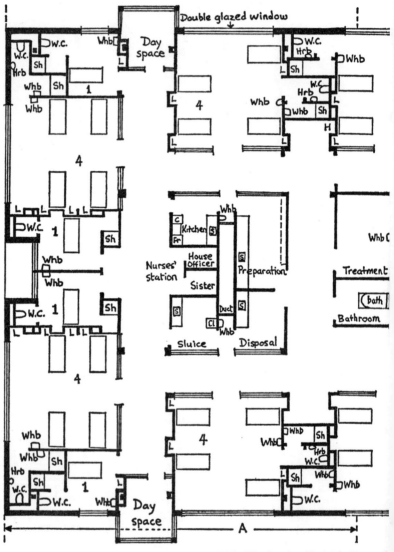

Fig. 155. The Falkirk Ward unit. (*Scottish Home &*
The area A of the plan is

1..single bed ward   CI..clinimatic disposal unit
4..four bed ward   Fr..fridge
C..cooker   H..hose reel

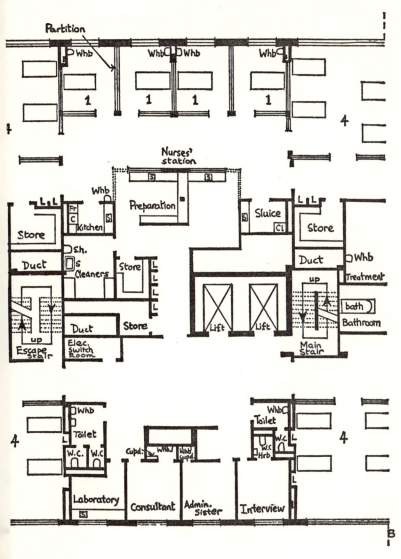

Health Department and H.M.S.O., Crown copyright)
reproduced, in reverse, beyond B.

Hrb..hand rinse bowl   Sh..shower
L..patient's locker   S.h...slop hopper
S..sink   Whb..wash hand basin

The commonest type of ward until quite recently was the 'open ward', i.e. one accommodating some thirty patients (sometimes as many as fifty) with beds set at right angles to the walls; by the use of screens, small partitioning, etc., other and more attractive arrangements of the beds was possible (Fig. 154).

In recent years, particularly since the end of the Second World War, much thought has been given to ward design, resulting in the following features:

(1) The subdivision of beds into small rooms, usually of four, two and one bed each. Such an arrangement has many advantages; it is possible for patients to have greater privacy, patients may be allocated accommodation according to their clinical and nursing needs, e.g. a seriously ill patient would be placed in a single ward. This arrangement also makes possible greater flexibility in use, e.g. the number of male or female beds may be varied according to the demand.

(2) Careful siting of nurses' stations. This facilitates adequate observation of patients, particularly patients who are seriously ill.

(3) Grouping of nurses' duty rooms, treatment rooms, sluice rooms and toilets, etc. This arrangement makes possible a more efficient planning of the engineering services, an important element in the cost of ward construction.

Take Fig. 155 as an example of such a ward which has been constructed in Falkirk in Scotland. It should also be noted that the arrangements result in patient accommodation being placed on the outside of the building, thus providing light, ventilation and a much more pleasant outlook.

Another development is the concept of the 'intensive care unit'. This, briefly, is based on the principle that only a certain proportion of patients require continued observation and attendance of nurses and doctors. These patients are accommodated in a special ward where all the aids to treatment, e.g. oxygen supply, suction, monitoring apparatus, are available. Staff can be concentrated on such patients, the remainder of the ward not requiring to be so heavily staffed.

Ventilation

The arrangement of beds in the ward is such that each patient has 1,200 ft³ of air, and the ventilation must be arranged so that this air may be changed at least three times in the hour. The floor space for each patient must be at least 100 ft², and the height of the ward is 12 ft. Any height above this is uneconomical, being of no value for ventilation. This is because the warm moist air rises and forms a belt at this level. The beds should be placed at intervals of 5 or 6 ft apart, and usually they are arranged alternately with the windows. Experiments are, however, being made in arranging beds longitudinally to give the patients a better outlook from the windows, and to avoid what is known as 'droplet' infection. (This is the theory that when a patient has an infection such as a cold or tuberculosis the germ is spread by droplets of saliva which are sprayed from the patient's respiratory passages as he coughs and sneezes or even talks vigorously.) The window space per patient to allow for adequate light must be at least 24 in^2, but is usually much more. The most popular windows are of the Hopper variety (see page 276) because it is essential that the air should be directed upwards to avoid draughts. The Hinke's Bird method of ventilation is very useful in hospital wards which have sash windows (see page 276). In ventilating the wards attention should be paid to the prevailing winds, so that if there is a keen north wind blowing, the windows on the south side should be the ones to be opened, but this naturally depends on the aspect of the ward. Windows opposite each other should not be open together as this creates a draught. The thermometer should be consulted at intervals during the day, and an average temperature maintained of 60°F. Where thorough ventilation is difficult, screens are helpful in avoiding draughts. Other devices commonly used in ventilation of wards are inlets behind the radiators, about 6 in from the floor. These gratings admit cold air, which is warmed, and thus rises into the ward without causing draughts along the floor. In hot, still weather, the use of the electric fan is invaluable, especially for certain patients whose body temperature is difficult to regulate, and who are liable to suffer from heat retention.

Heating

The problems of heating hospital wards are similar to those in other buildings. The most common method is heating by steam radiators, generally placed under the windows. The advantages of these are their cleanliness and their efficient radiation, which gives a well distributed heat to the ward. They can be easily regulated by a stopcock, and last but not least of their advantages, from the nurse's point of view, is their usefulness in heating and airing blankets and clothing. Panel heating has been introduced in some of the newly built hospitals. The pipes are hidden in the walls or ceilings and with proper regulation the method is claimed to be a success.

Electric radiators are very efficient but expensive to run. The electric elements may be built into walls and ceilings as in panel heating.

The problems of heating and ventilating hospital wards and theatres are often solved in modern hospitals by what is known as 'air conditioning' – the air is pumped into the building by some mechanical means (see page 278), but before it reaches the ward or theatre it is conditioned, i.e. warmed, filtered, dried or moistened and rate of entry regulated. When this method is in use the exits for the air after it has circulated have to be carefully placed in relation to the inlets.

Lighting

Indirect lighting, i.e. by light reflected from the ceiling, is the most restful. It is now usual to provide individual bed lights for each bed, and the switch panel also allows for wireless installations, alarm bells, etc. In more modern wards night-lights, suitably shaded, are fitted into the walls near to the floors. They do not disturb patients but enable the work of the ward to be safely carried out.

In addition to having artificial lighting it is of course essential that the ward should be bright and cheerful, with large windows. In operating theatres much consideration is given to efficiency of the lighting, and large 'shadowless' lamps are usually fixed. A modern development is to have dome-shaped theatres provided with multiple lighting points, so that all or any of them can be brought into use.

Sanitation and disposal of refuse

The waste from a hospital ward may be divided into three types:

(1) Dry refuse.
(2) Waste water.
(3) Human excreta.

(1) **Dry refuse,** i.e. the contents of pails, dressings, food, etc. This is usually disposed of by being emptied into an incinerator. Very often these incinerators are part of the heating systems for the boilers in the Hospital Engineer's Department. The stokers rake the refuse on to the furnaces. Unburnable refuse, after it has been through the fire, is

removed as 'clinker' and is sold for use in making macadamized roads.

(2) **Waste water,** i.e. bath and sink water, leaves the building by the usual waste-water pipes, which are interrupted at intervals by U-bends, gully-traps, etc., and finds its way into the main drains.

(3) **Human excreta** is emptied in a special room called the 'sluice-room'. Here special arrangements are made for the thorough emptying and flushing of the bed-pans and there is apparatus provided for their steriliza-tion also. The excreta leaves by the 'soil-pipe' affixed to an outside wall. It goes straight underground to the sewer via the manholes, and is ventilated at its upper end above the roof.

Hospitals in the country usually have their own sewage works for disposal of sewage on general lines (see Chapter 7).

Other points of practical interest in connection with hospital buildings are:

(1) The floors should be of hard, non-absorbent material, which does not stain easily. Different forms of composition or marble or wooden blocks are commonly seen.

(2) All corners and junctions are rounded to facilitate removal of dirt.

(3) Doors are plain with no ledges and are generally fitted with springs to avoid banging.

(4) Various kinds of taps are fitted on to wash basins. They may be operated by the elbow or foot, so that the surgeon or nurse, having scrubbed the hands, does not need to touch the taps with them.

The sick-room in a private house

It is often necessary to nurse a patient at home, and if the illness is a severe or long one, it is well worth while giving thought to the room to be chosen for such a purpose. The position is of importance: it should have southerly aspect if possible, to get maximum light and sun, but blinds or dark curtains are a necessity to give a restful atmosphere when the patient requires sleep, and the bed should be so placed that strong light does not shine into the patient's eyes. The room should be in a quiet part of the house, near a bathroom and lavatory, and not too far away from the kitchen. The general principles of heating, lighting and ventilation must be applied. If the room is small and difficult to ventilate, a screen should be put round the bed, and the windows and door opened at intervals to ensure fresh supplies of good air.

Infectious diseases and tuberculosis hospitals

Special precautions are necessary in hospitals dealing with infectious diseases. The bed spacing is more generous than in general hospitals and this reduces the risk to the other patients and staff of droplet infection. In intestinal infectious diseases special care has to be taken in the disposal of the excreta. The success of preventive measures depends essentially on the nurses observing the rigid procedure which is laid down. Such care is, of course, necessary in general hospitals, but all the more so in hospitals dealing with infectious diseases.

The same remarks apply to the care of patients suffering from tuberculosis, which is, of course, also an infectious disease. In pulmonary tuberculosis, the patient's sputum is highly infectious. Disposable containers which can be made of cheap material such as cardboard and which can be destroyed after use, are the most satisfactory means of receiving the sputum. Permanent containers require to be sterilized after use. The cleansing or disposal of sputum containers should be undertaken by staff who realize the dangerous risk of infection from sputum. Training in personal hygiene is essential for patients suffering from pulmonary tuberculosis, and the maintenance of such hygienic standards is essential when the patient returns to his own home.

The outlook regarding tuberculosis has changed dramatically in recent years. The advent of antibiotics and chemo-therapeutic drugs has revolutionized its treatment, and it now can be claimed that the majority of cases are curable. Due to the use of these drugs which amongst other things render the patient non-infectious, and the work of the public health authorities in tracing contacts of cases and ensuring that they obtain early treatment, the incidence of the disease has greatly diminished. The requirement of beds for such patients has also been greatly reduced, and many tuberculosis hospitals are now being used for other purposes. It must be emphasized that the control of tuberculosis depends on continued effort of hospital physicians, public health medical officers, district health visitors and others, and their efforts can in no way be relaxed. Tuberculosis is still a very acute problem in the developing countries.

Although epidemic diseases are still of importance to the health of the community, as with tuberculosis, there has been a great reduction in their incidence in the last 25 years; the hospital beds no longer required to be retained for infectious diseases have been used for other purposes; the term 'isolation' or 'infectious diseases' hospital is really no longer appropriate, and these hospitals are now dealing with many general, medical and surgical conditions.

Hospital diets

The question of diet should be of particular interest to the nurse, for without good food the healing processes of the body are retarded. The ordering of meals is not in the junior nurse's province, but as she proceeds to posts of responsibility she consults with physician, kitchen superintendent, stores manager and cook to convert the theory of dietetics into the practice of actually feeding the patient. Hence the importance of knowing food values and the varieties suitable for different conditions. She must always remember that the nurse is the person who finally presents the food to the patient, and it is useless to have correctly chosen and well-cooked food unless the service is perfect in every way, i.e. dainty trays, shining crockery, well polished silver and small helpings attractively arranged and served really hot or cold as required. It is of course a great advantage to a nurse if she can cook simple dishes. A course of practical cookery should be taken, along with the study of balanced diets, before the student enters hospital. The Ministry of Health and the Ministry of Food keep a strict eye on the purity of the milk supply in this country, and when nursing the sick, Tuberculin tested milk should be obtained. In the case of babies, milk even of the highest grade is boiled for 5 minutes, as the mucous membrane of the alimentary tract of the young child has not gained its immunity to infection. No risk must be run, especially when the child is sick, of adding to its difficulties.

Children's wards

It is now generally accepted that when children require hospital treatment they should be accommodated in wards or units (and preferably hospitals) of their own, and it is not good practice to deal with children in adult wards except when this is unavoidable, e.g. in cases of emergency or when highly specialized facilities are available only in adult hospitals. It is particularly important, in the care of sick children to ensure that everything is being done to alleviate mental as well as physical suffering.

The chief thing to keep in mind in the planning of wards for sick children is the possibility of 'cross-infection'. The immunity of young children to infection is only developed with years, or by suffering attacks of the various diseases. When children are ill from almost any cause, their immunity is low, and they are extremely liable to contract other infections with which they may be in contact. Consequently outbreaks of measles, chicken pox, epidemic diarrhoea, etc., have been a great draw-back to the nursing of children in general wards. This has led to the modern development of cubicle nursing, and the use of observation wards. Into the latter the children are put for 24 to 48 hours until their condition is diagnosed, so that if anything of an infectious nature is discovered, the child can be nursed in isolation. The cubicles may each contain one, two or three cots and the partitions are built of unbreakable glass. Hence the nurses as they pass can see the children without entering the cubicles and much cross-infection is avoided. Each cubicle contains all that is required for nursing the children, i.e. separate wash basins, bath, bath apron, thermometer and overall for the nurse. In addition to the ordinary annexes there should be attached to a children's ward a large airy nursery or play-room, with little chairs and tables and suitable toys. The disinfection and care of toys presents a problem, and the nurses must be constantly on the watch for any toys which may become infected or dirty. They must be frequently sorted out and unsuitable ones burnt. It is of course essential that convalescent and ill children should be allowed toys to play with. There is no sadder sight than to see children sitting up in bed unoccupied because 'toys make the ward untidy'. Fortunately there are few hospitals where such an idea would be tolerated nowadays. 'Occupational therapy' has become a part of the treatment of disease, and special teachers are often employed to teach children (and adults too), who are in hospital for long periods, some craft or occupation.

A children's ward always has a room set apart for the preparation of feeds. It is called the milk kitchen, and there all bottles are kept and sterilized, and the milk is kept in a refrigerator. Nurses working amongst sick children must realize above all the possibility of spread of infection, and to prevent this their own personal hygiene must be beyond reproach. They must keep their nails short and clean, and must wash their hands thoroughly after attending to each baby and before going to the next. Nothing is more dangerous than for a nurse to give nursing attention to a child and then to feed him before thoroughly scrubbing and disinfecting her hands. The infant's delicate alimentary tract is not proof against infection so carried. When the nurse has to carry a child in her arms she must let him face over her shoulder so that droplet infection is not carried from one to the other. The nurse has her own health to consider too, and should not put herself in a position in which the child can cough in her face. When feeding a child in bed, she must remember this and sit at the side rather than in front of him. Nurses with heavy colds are usually excluded from children's wards, but if it is absolutely necessary for them to be amongst sick children they should wear a gauze mask over the mouth and nose.

The architecture of hospitals is constantly changing in its main outlines and practical details, but any serious student of hygiene will find that the principles she has learnt are being increasingly applied, in order to make the surroundings of the patient more satisfactory in every way.

11 Public administration with regard to health matters

The State has shown an interest, if only an erratic one, in the health of its citizens since ancient times. The Black Death produced legislation to meet the situations which arose. The quarantine regulations for ships are another example of health legislation in early times. Ships from places where plague existed were held up for a period of forty days – hence the name 'quarantine'.

Organized public health in England and Wales is a product of the nineteenth and twentieth centuries. The immediate cause of the setting up of the first organized health services was an outbreak of cholera in 1831, when a consultative Central Board of Health was appointed to advise on the steps to be taken to combat this disease. This 'Central Board of Health' was appointed to deal with an emergency, but from it has developed the present-day health organization.

The social changes in this country, due to the industrial revolution, brought with them much squalor. Housing conditions were wretched and good water supplies were a rarity – water-carriage sanitation was almost unknown. The first duty of the hygienist was therefore to improve the environment in which the masses of the people lived, and the nineteenth century was concerned with environmental hygiene – housing, sanitation, water supply, etc.

The twentieth century saw a continuation in the improvement in the environmental conditions of the population. The problems of working-class housing and slum clearance were attacked. A new conception – a conception that the State is concerned with the health and fitness of the individual – was born; out of this idea the personal health services such as the School Medical Service, Tuberculosis Service, Maternity and Child Welfare Schemes, etc. were developed.

With the passing of the National Health Service Act 1946 and National Health Service (Scotland) Act 1947, the State accepted responsibility for the provision of a Health Service dealing with both curative and preventive medicine free of charge to those requiring it. The cost of the service is met partly by compulsory contributions out of salaries and wages and partly out of taxation and rates.

Although the State has made itself responsible for the provision of the Health Services, such services are not administered entirely by Governmental Departments and the State still looks, in large measure, to Local Health Authorities for the provision of many of the services.

The Local Health Authorities are:

(1) **County Boroughs.** These are the larger towns. Any town wishing to become a County Borough must now have a population of over 75,000. County Boroughs are autonomous units of local administration, responsible for all health service in their area, and subject only to the general supervision of Central Government departments.

(2) **Administrative Counties.** The Administrative County corresponds in area, generally speaking, to the geographical counties. A few of the geographical counties are divided into more than one administrative county, e.g. the three Ridings of Yorkshire.

The Counties administer directly within their own area most of the personal hygiene services, such as the Tuberculosis Schemes, the Midwives' Act and certain quasi-hygienic Acts such as the River Pollution Prevention Act, the Rats and Mice Destruction Act; they usually administer the Food and Drugs Act, and Maternity and Child Welfare Schemes.

The County Councils are the supervising and co-ordinating authorities over the County Districts which make up the administrative county. They have no jurisdiction in County Boroughs.

(3) **Municipal Boroughs, Urban Districts and Rural Districts** form the area of the administrative county. These authorities are 'sanitary authorities', and are responsible for environmental hygiene, e.g. water supplies, sewage, scavenging, etc., cleanliness of cow-sheds. The health functions of the Municipal Boroughs and larger Urban Districts approximate to those of the County Borough and include some of the personal health services.

M

In smaller urban areas and rural districts the County Council is responsible for most health services.

The Central Government department mainly responsible for the
supervision of the health services is the Ministry of Health. The Ministry
of Health exercises a general control but does not interfere in day-to-day
administration.

In Scotland the system is similar although the nomenclature is different.
Counties, Counties of Cities and large Burghs correspond, from the point
of view of the Health Service, to Counties and County Boroughs. Small
Burghs are similar to the Municipal Boroughs. There are no Urban Dis-
trict Councils and Rural District Councils in Scotland, their function
being undertaken by the County Councils themselves. The Scottish Home
Health Department exercises the general control, as does the Ministry of
Health in England and Wales.

The National Health Service

As already stated, the National Health Service includes in its scope both
the prevention and the after-care of illness and treatment. The Government departments concerned are the Ministry of Health in England and
Wales and the Scottish Home and Health Department. There are three
branches of the Health Services:

(1) Hospital and Specialist Services

The responsibility for the provision of this service is placed on the Ministry
of Health and the Secretary of State for Scotland, and the service is, for
all practical purposes, financed out of Exchequer funds.

On the passing of the National Health Service Act, all hospitals,
whether voluntarily run by their own Boards of Governors, or hospitals
belonging to Local Authorities, were taken over by the State. For
hospital purposes, the United Kingdom is divided into 20 regions, 15 in
England and Wales and 5 in Scotland, each one usually based on, or
associated with, a medical school.

The responsibility for administration of the Hospital Service is vested
in a Regional Hospital Board. These Boards are composed of persons
interested in hospitals, nominated by the Ministry of Health or the Secretary of State. They are unpaid.

The main function of the Regional Hospital Boards is the overall
administration and the planning and development of hospital services.
The day-to-day running of hospitals is carried out by Hospital Manage-

ment Committees or Boards of Management appointed by the Regional Board.

In addition to their responsibility for the provision of hospital treatment for patients, hospital authorities also hold the important administrative responsibilities connected with the training of medical students and nurses. They ensure that all reasonable facilities are available to medical schools for the education of their students. Most of the larger hospitals, either by themselves, or in co-operation with other hospitals, have training schools for nurses at which they can be qualified.

It should be appreciated that hospitals demand the services of persons of many skills and, in addition to doctors and nurses, hospital authorities are concerned with the training of physiotherapists, occupational therapists, almoners, administrators, etc.

(2) General Practitioner Service

Every person in the United Kingdom who has joined the National Health Service is entitled, free of charge, to the services of a general practitioner or 'family doctor' of his choice. Having chosen a doctor and been accepted by him the patient goes on the 'doctor's list' and the medical practitioner receives payment from State sources for the patients on his list. Practitioners taking part in the 'General Practitioners Service' are able to treat patients under private arrangements, if the patient so wishes; the practitioner cannot, however, treat a patient on his list as a private patient.

The general practitioner scheme is administered by a body known as an 'Executive Council' the members of which are appointed by the Minister of Health, or Secretary of State for Scotland. Usually, but not always, the area of an Executive Council is that of a County Borough or County Council.

(3) Preventive Service

The preventive and after-care services are administered by the Local Health Authorities. The purpose of such services is to prevent serious illness developing and to assist anyone who has been ill to re-establish his, or her, place in society. These services are extensive and deal with such matters as ante-natal, maternity and child welfare care, immunization, the care of the physically and mentally handicapped and old people, etc. They are paid for by the Local Authority. Most services attract a Government grant which is payable when schemes are approved. This Exchequer grant enables the Central Government to exercise a general control over the Local Health Authorities' services.

(4) Voluntary bodies

Although the National Health Service provides for the prevention, treatment and after-care of illness, there is still much scope for voluntary effort in connection with the health services. Voluntary societies exist to help special classes of individual, such as epileptics, spastic children, etc., and most cities have voluntary bodies interested in the welfare of the aged and chronic sick, physically and mentally handicapped, etc. In the case of some hospitals, private persons form themselves into associations with titles such as 'League of Friends', their aim being to interest themselves in the work of the hospital and to provide 'extras' for patients which cannot be provided from official sources.

The National Health Service Act aims at providing a health service covering the whole of the population. In addition to the facilities provided under this Act, there are numerous other organizations providing health services for special groups of the population. Using the term in its very widest sense, it is now generally accepted that the 'good employer' should take an interest in the health and welfare of persons employed or those with which an organization is concerned. For example, the fighting services are responsible for the health of their own personnel and special services are available to merchant seamen, those working in the mines, etc. It is also usual for large industrial firms to provide a medical service for the workers and for universities' special services to supervise the health of students. Reference is now made to such services:

(1) The School Medical Service

The Service is provided under the Education Act. The origin of this service arose from the fact that, when education was made free and compulsory, it was found that a considerable number of children were not able to benefit fully from their education owing to physical and mental defects.

The School Medical Service, which is supervised by the Department of Education and Science, is run by the Local Education Authorities of County and Borough Councils. The Service provides for the examination of school children to detect defects, and makes arrangements to provide necessary treatment or special care for children who are found to be physically or mentally handicapped.

The School Medical Service must work in close co-operation with other health services, the general practitioner services and hospital service. It has particularly close association with the Local Health Authority Service, the Medical Officer of Health, usually being also the School Medical Officer.

(2) Services under the Factory Act

The industrial revolution in the early nineteenth century led to a great multiplication of factories throughout the country, and unfortunately the conditions in many factories were extremely poor – one particular point which stirred the public conscience was the poor conditions in which many children of very tender years had to work for very long hours. The State has now accepted the responsibility for ensuring that reasonable conditions are provided in the factory. Child labour is no longer permitted.

The first Act of Parliament dealing with the conditions of factories was passed early in the nineteenth century and since then there has been a series of factory acts, the most recent one being 'Factory Act 1961' which consolidates and amends previous Acts.

The Factory Acts deal with such matters as maximum hours of work, general conditions of work, health hazards and the prevention of accidents. Welfare provision, first aid and medical provision – the particular requirements in regard to health in various industrial processes are dealt with.

The administration of the Factory Act is the responsibility of the Ministry of Labour and National Service. This Department provides a factory inspectorate staffed by H.M. Inspectors of Factories. They are responsible for visiting factories to advise on measures to comply with the Act, and generally to assure themselves that all provisions are being observed. The Factory Inspectorate includes a medical branch concerned more particularly with the health hazards in factories.

It is clear that the State has now accepted very fully its responsibility for the health of its people. The statutory provision has been briefly referred to. Reference to health and welfare services would be incomplete without making mention of the major contributions made by voluntary societies and associations interested in particular aspects. There are for example a number of voluntary associations whose function is designed to help those suffering from mental ill-health.

Vital statistics

Vital statistics give an index, by means of figures, of the state of the population. They supply a method of measuring the hygienic state of a population, and enable a national 'stock-taking' to be made.

Population

The basic figure in all work on vital statistics is the number of inhabitants of an area, i.e. its population. This figure is obtained by means of the census, which is taken in Great Britain every ten years. A householder is required to supply information regarding sex, age, whether married or single, and occupation, of all persons in his household.

The data obtained from the census is analysed by the Registrar-General's Department. Amongst other information this department supplies the population of each Local Authority area.

The population, based on a census, is only correct for the year in which the census is taken. The population of intercensal periods may be calculated by assuming the increase of population in any intercensal period is at the same rate as in the previous intercensal period.

Death rate

Before the disposal of a body a death has to be certified by a medical practitioner. If, for any reason, such a certificate is not forthcoming, a coroner's order is required before the disposal of a body. It is thus possible to collect accurate information of the number of deaths and the various causes of death.

For purposes of comparison between two districts it is not sufficient to know the total number of deaths in the two districts. Twenty deaths in a village of 500 inhabitants is a more serious affair than 20 deaths in a village of 3,000. For comparison with other districts the deaths in any area are expressed as the number of deaths per every thousand of the population. This is termed the 'death rate' per 1,000 population.

The 'death rate' per 1,000 population is calculated as follows:

Let p = population of area, d = total deaths per annum in the area. Then death rate per annum per 1,000 population

$$= \frac{1,000}{p} \times d.$$

The 'death rate' gives some index of the healthiness of an area; 'other things being equal' an area with a high death rate is unhealthier than that with a low rate.

'Other things being equal' is a phrase of some importance. Some towns, e.g. Cheltenham, Bath, have a large population of retired elderly people, while new manufacturing towns have a preponderance of young people. Quite apart from the 'healthiness' of the area, Bath would show a larger death rate than a new manufacturing district. The 'death rate' calculated as already explained is the crude death rate. In order to make the comparison between areas a true one it is possible to calculate a death rate from the crude death rate on the assumption that the population of a town has the same age distribution as the general population. Such a death rate is the 'standardized death rate'.

Marriage rate

Marriages are registered as are deaths, and the marriage rate per annum is:

$$\frac{\text{number of marriages per annum} \times 1,000}{\text{population}}$$

A high marriage rate is said to be a sign of prosperity in a district.

Birth rate

This is expressed in the same manner as the death rate.

$$\text{Birth rate} = \frac{\text{number of births per year} \times 1,000}{\text{population}}$$

This rate is not a true index of fertility. The number of children depends on the number of married women of child-bearing age.

The fertility rate is:

$$\frac{\text{number of legitimate births} \times 1,000}{\text{number of married women between 15–45 years}}$$

and the illegitimate birth rate:

$$\frac{\text{number of illegitimate births} \times 1,000}{\text{number of unmarried women and widows between 15–45 years}}$$

Notification of diseases

The only diseases about which we have accurate numerical information are the notifiable diseases. This information is as accurate as the diagnosis of the disease. The diseases are the infectious diseases made notifiable by the Infectious Diseases (Notification) Act 1880, now incorporated into the Public Health Act 1936.

The diseases originally mentioned were smallpox, cholera, diphtheria, membranous croup, erysipelas, scarlet fever, typhus, enteric fever, relapsing fever, continued fever, puerperal fever. It will be noticed that some terms used, e.g. 'continued fever', are old-fashioned and are hardly ever now used in medicine.

The notifications are made to the Medical Officer of Health of the Local Health Authority, and the Act gives power for diseases to be added to the Act. Other diseases have been added, e.g. plague, cerebrospinal fever, tuberculosis. Measles and whooping-cough were added to the list in October, 1939, as it was thought such a step would help to combat these diseases in the population evacuated on account of the war. It should be noticed that some infectious diseases are not generally notifiable, e.g. chicken-pox and German measles.

Information about diseases can be expressed statistically in several ways:

(1) Morbidity rate indicates the number of persons affected by a disease per 1,000 population.

$$\text{Morbidity rate} = \frac{\text{number of cases} \times 1,000}{\text{population}}$$

(2) The mortality rate indicates the number of persons who *died* of a disease, e.g. tuberculosis, per 1,000 population.

The infant mortality rate is the number of deaths of infants under one year per 1,000 registered live births.

It is said that the infant mortality rate, the tuberculosis mortality rate, and the standardized death rate are the most sensitive indices of social conditions.

Statistics may be collected in connection with the many activities of health departments, hospitals, etc. A number of routine returns give useful information and statistics have also their part in special inquiries. They are of great assistance in assessing the work of any service and planning its developments.

The place of such statistics in any investigation, especially their limitations, must be appreciated. Lack of this appreciation gives rise to the remark that 'figures prove anything'. It must be remembered that the manipulation, interpretation and presentation of statistics require considerable experience.

Morbidity rates and mortality rates of a disease must not be confused with the case mortality, or fatality rate of the disease.

The case mortality of any disease is the percentage of cases suffering from the diseases which resulted in death. This figure is a *percentage* and it is the only rate *not* calculated per 1,000 population.

Where c = number of cases, and d = number of cases resulting in death.

The case mortality is $\dfrac{d}{c} \times 100$

An interesting development in health and hospital administration is the use of computers in the processing and analysis of statistics. One of the difficulties is that the results of any statistical investigation may not be available for some considerable time after the basic data has been obtained – too late for effective action to be taken if it is necessary. With the use of computers, the results of a statistical investigation can be made available after a much shorter interval, and appropriate action can be taken without the former delay.

It should be emphasized that, where statistics are used, the purpose of investigation must be clearly understood. Failure to appreciate this very obvious point brings statistical methods into disrepute.

World Health Organisation

In 1964 the World Health Organization (W.H.O.) with permanent headquarters at Geneva was founded under the auspices of the United Nations.

Organization co-ordinates research into diseases throughout the world and where the need is great makes funds available for the distribution and administration of drugs used in the campaign against such diseases as malaria, tuberculosis, smallpox and yaws. It also plays a large part in health education and propaganda.

Index

Fibrous tissue, 14
Fibula, 137
Filiform papillae, 182
Filium terminale, 156
Filter-bed, 294
Filtration: by kidney, 145
  of water, 294–5
Finger: muscles, 59
  prints, 139
Fires: open, 274, 276, 281
  gas, 282
First filial generation (F$_1$), 210–13, 216
Fish (diet), 265
Fission (bacteria and protozoa), 312
Fissures of cerebrum, 161
Fixation muscles, 47
Flagella, 8
Flannelette, 246–7
Flatworms, 325–7
Fleas, 316, 321–2
Flexion (muscular), 48
Flexor muscles, 58, 59, 61–2
Flies, 300, 315, 327–30
Floors, 287, 322, 339
Floor space, 275
Flour (diet), 265–6
Flukes, 325
Fluorescent lighting, 285
Fluorine, 244, 258
Focusing, 176–8
Foetus, 200
Follicular epithelium, 194
Follicle stimulating hormone (FSH), 203
Fomites, 316, 317
Food, 110–15, 252–63
  preservation, 270–1, 327–8
  poisoning, 271–3
  vacuole, 5
Foodstuffs, 264–7
Foot, 37–8
  movement of, 44
Foot-candle, 284
Foramen magnum, 40, 160–1
Foramina (skull), 25
Forearm, 45
Fore-brain, 138, 166–7
Formaldehyde, 319
Foundations of buildings, 286
Frontal bone, 42, 99
Fructose (fruit sugar), 110–11, 135
Fruit fly, 214
Fuel, 281–3

Fumigation, 319, 322, 323
Fundus: of bladder, 148
  of stomach, 122
  of uterus, 196
Fungiform papillae, 182
Fur, 153

Gall bladder, 28, 124–6, 131, 132, 136
Gametes, 210–13, 215–17
Ganglion, 32
Gas disinfection, 319
Gas fires, 282
Gastric: factor, 124
  glands, 258
  juice, 123, 124
Gastrocnemius, 61–2
Gastrocolic reflex, 128
General practitioner, 347
Genes, 212, 213, 217
  hidden, 216
  structure of, 236
Genotype, 213, 215
Germ cells, 187, 190, 194, 212
German measles, 352
Giddiness, 181
Gigantism, 208
Glands: adrenal, 207
  Brunner's, 125
  ductless, 13, 205–9
  gastric, 123
  Lieberkuhn's, 125
  lymph, 125
  parathyroid, 206
  pituitary, 162, 203, 208–9
  prostate, 192, 193, 202
  salivary, 119
  suprarenal, 207
  thymus, 206–7
  thyroid, 205–6, 258
Glandular epithelium, 12
Glenoid cavity, 34
Gliadin, 112, 266
Globulins, 112, 127
Glossopharyngeal nerve, 165
Glottis, 99, 102, 121, 124
Glucose, 110–12, 135, 137
  in blood, 65, 137–8, 148, 208
  formula of, 226
  phosphate, 237
Glutamine and glutamic acid, 229
Gluteal muscles and nerve, 61
Gluteins, 112

Maxillary nerve, 164
McKinnel's ventilator, 277
Measles, 312, 313, 314, 315, 342, 352
'Measly pork', 326
Meat (diet), 265, 270
Meatus (auditory), 41, 178
Mechanical ventilation, 278
Median: nerve, 158
  vein, 83
Mediastinum, 28, 100
Medical: Health Officer, 349
  training, 347
Medium (organism culture), 309
Medulla: of adrenals, 207
  of kidney, 143
Medullary sheath, 19
Meiosis, 189–90, 217
Meissner's corpuscles, 184
Membrana: granulosa, 194, 195
  tectoria, 180
Membranes of cell, 221, 222, 223
Membranous: croup, 352
  labyrinth, 180
Mendel, Gregor: laws of, 211, 213
  work of, 210–14
Meninges, 155–6
Meningitis, 312
Menopause, 200
Menstruation, 196, 198–200, 203
Mercury salts, 319
Mesenteric: arteries, 79, 124
  veins, 85
Mesentery, 30, 124
Mesogloea, 6, 7, 8
Messenger RNA, 235
Metabolism, 107
  basal, 253–5
  regulation, 207
Metacarpal bones, 34
Metatarsal bones, 37
Methane structure, 224
Methionine, 229
Methylene blue (milk test), 269
Mid-brain, 161, 167
Micro-organisms, 267, 274, 295, 308–20
Microscope: electron, 220
  optical (light), 219, 220
Micturition, 149–50, 315
Middle ear, 179
Milk, 197, 198, 256, 258, 266–70, 315, 341
  composition of, 266

kitchen, 342
sugar, 110–11
Millon's test, 113
Mineral salts, 139, 142, 257–8
Miner's cramp, 147
Ministry of Agriculture and Fisheries, 267
Mites, 324
Mitochondria, 221, 222, 223, 236, 238–9
Mitosis, 188–9
Mitral valve, 75
Modiolus, 180
Molar teeth, 119
Molecular: biology, 223
  models, 223–6, 230–1, 234
Monocytes, 68
Monoecious (tapeworm), 326
Monosaccharides, 110, 225–6, 227
Mons pubis, 197
Morbidity rate, 352
Mortality rate, 352, 353
Morula, 200
Mosquitoes, 328–9
Motor nerves, 157, 163
Mountain sickness, 106
Mouth, 116, 244
Movement, 4, 8
  of air, 275
  of joints, 43–4
Mucin, 119, 126, 128
Mucous membrane, 260
  of alimentary canal, 117, 120–1, 123, 181,
  of ear, 179, 181
  of eyelid, 171
  of respiratory organs, 29, 42, 98–9, 108, 183
  of uterus, 200
  of vagina, 196
Multicellular animals, 6
Munro, foramina of, 162
Muscle: action, 6, 8, 46–7
  sense, 185
  structure, 17–18, 19–20
  tone, 151, 153
Muscular system, 46–63
  of Hydra, 8
Muscular tissue, 17
Musculo-cutaneous nerve, 158
Musculo-epithelial cells, 6, 8
Musculo-spiralis nerve, 158
Myocardium, 74

Woollen clothing, 247
World Health Organization, 353
Worms, 325–7
Wrist, 34
    muscles of, 59

X and Y chromosomes, 190
Xanthoproteic test, 113
Xenopsylla cheopsis (rat flea), 322
X-ray analysis, 220, 230, 233

Yaws, 353
Yeast, 261, 266
Yellow elastic tissue, 14
Yellow spot, 173
Yolk sac, 200
Young–Helmholtz Theory, 176

Zein, 112
Zona pellucida, 195
Zygomatic bone, 41, 42
Zygote, 187